Cost/Price Analysis:
Tools to Improve
Profit Margins

Cost/Price Analysis: Tools to Improve Profit Margins

LeRoy H. Graw
C.P.M., CPCM, EdD

VAN NOSTRAND REINHOLD
_____ New York

Copyright © 1994 by Van Nostrand Reinhold

Library of Congress Catalog Card Number 93-12856
ISBN 0-442-01717-0

All rights reserved. No part of this work covered by
the copyright hereon may be reproduced or used in any
form or by any means—graphic, electronic, or
mechanical, including photocopying, recording, taping,
or information storage and retrieval systems—without
the written permission of the publisher.

I(T)P Van Nostrand Reinhold is an International Thomson Publishing company.
ITP logo is a trademark under license.

Printed in the United States of America.

Van Nostrand Reinhold
115 Fifth Avenue
New York, New York 10003

International Thomson Publishing
Berkshire House, 168-173
High Holborn
London WC1V7AA
England

Thomas Nelson Australia
102 Dodds Street
South Melbourne 3205
Victoria, Australia

Nelson Canada
1120 Birchmount Road
Scarborough, Ontario
M1K 5G4, Canada

International Thomson Publishing GmbH
Königswinterer Str. 518
5300 Bonn 3
Germany

International Thomson Publishing Asia
38 Kim Tian Rd., #0105
Kim Tian Plaza
Singapore 0316

International Thomson Publishing Japan
Kyowa Building, 3F
2-2-1 Hirakawacho
Chiyada-Ku, Tokyo 102
Japan

16 15 14 13 12 11 10 9 8 7 6 5 4 3 2 1

Library of Congress Cataloging-in-Publication Data

Graw, LeRoy H.
   Cost/price analysis : tools to improve profit margins / LeRoy H.
Graw
      p.   cm.
   Includes bibliographical references and index.
   ISBN 0-442-01717-0
   1. Industrial procurement—Cost control.   2. Costs. Industrial.
3. Prices.   4. Profit.   I. Title.
HD39.5.G73   1993
658.15'5—dc20   93-12856
                                                                                CIP

To my wife, Anat, and to our children,
Byron and Karen.

Contents

Foreword   xiii

Chapter 1   **Introduction**   **1**
PURPOSE OF THIS BOOK   1
THE BUYER'S PRICING TEAM AND THE
    RESPECTIVE ROLES AND RESPONSIBILITIES
    OF TEAM MEMBERS   1
DEFINITIONS   2
COMPARISON OF DIFFERENT ANALYTICAL
    METHODS FROM THE VIEWPOINT OF
    COMPLEXITY AND QUANTITY OF
    RESOURCES REQUIRED   4
TYPES OF MARKETS AND SUPPLIER PRICING
    STRATEGIES AND THE ANALYTICAL
    METHODS APPROPRIATE TO EACH   5
SUMMARY   7

Chapter 2   **Types of Contracts and the Analytical Methods
Appropriate to Each**   **9**
INTRODUCTION   9
THE EFFECT OF RISK ON SELECTING
    CONTRACT TYPE   10

OTHER FACTORS INFLUENCING SELECTION OF CONTRACT TYPE 10
FIXED-PRICE FAMILY OF CONTRACTS 12
COST-REIMBURSEMENT FAMILY OF CONTRACTS 26
"HYBRID" CONTRACTS 30
AGREEMENTS AND PRECONTRACTUAL INSTRUMENTS 32
SUMMARY 32

Chapter 3 Price Comparison Methods and How to Use Them 33

DIFFERENT CONDITIONS REQUIRE DIFFERENT ANALYTICAL METHODS 33
PRICE ANALYSIS 33
PLACING OFFERS ON THE SAME COMPARATIVE BASIS 47
DISCOUNTS 49
TRANSPORTATION AND DELIVERY CONSIDERATIONS 52
LIMITATIONS OF PRICE ANALYSIS 53
SUMMARY 60

Chapter 4 Elements of Cost 63

DEFINITIONS: DIRECT AND INDIRECT COSTS 63
DEFINITIONS: VARIABLE AND FIXED COSTS 64
USING VARIABLE AND FIXED COST TO COMPUTE BREAK-EVEN VOLUME OR SALES PRICE 72
SUMMARY 76

Chapter 5 Estimating Cost and Obtaining Cost Proposals From Suppliers 79

WHY ESTIMATING IS IMPORTANT 79
METHODS OF ESTIMATING 79
DETAILED ESTIMATING EXAMPLE 82
ESTIMATING AND ITS RELATIONSHIP TO THE PURCHASING CYCLE 85

OBTAINING COST PROPOSALS (ESTIMATES)
FROM SUPPLIERS 86
SUMMARY 90

Chapter 6 **Performing Cost Analysis, Including Engineering Analysis and Accounting Analysis 95**

INTRODUCTION 95
USING THE PRICING TEAM 95
ENGINEERING (QUANTITATIVE)
ANALYSIS-LABOR 96
USING THE LEARNING CURVE 97
ENGINEERING (QUANTITATIVE)
ANALYSIS-MATERIAL 110
ENGINEERING (QUANTITATIVE) ANALYSIS:
OTHER DIRECT COSTS 110
ENGINEERING (QUANTITATIVE) ANALYSIS:
PROFIT OR FEE 111
ACCOUNTING ANALYSIS 111
ACCOUNTING ANALYSIS: LABOR RATES 111
ACCOUNTING ANALYSIS: MATERIAL PRICES 116
ACCOUNTING ANALYSIS:
OTHER DIRECT COSTS 118
ACCOUNTING ANALYSIS: OVERHEAD
AND G & A RATES 119
ACCOUNTING ANALYSIS: DEPRECIATION COST 123
SUMMARY 127

Chapter 7 **Assessing Risk and Developing Prenegotiation Profit Objectives 129**

DETERMINANTS OF FAIR AND REASONABLE
PROFIT OR FEE 129
METHODOLOGY FOR DEALING WITH PROFIT
OR FEE DETERMINANTS 129
SUMMARY 134

Chapter 8 **Conducting Price Analysis Supplemented by Partial Cost Analysis 137**

SITUATIONS CALLING FOR PRIMARY RELIANCE
ON PRICE ANALYSIS AND SECONDARY
RELIANCE ON COST ANALYSIS 137

PURCHASES FROM SUPPLIERS UNWILLING (OR UNABLE) TO PROVIDE COMPLETE COST DETAIL 143
SUMMARY 144

Chapter 9 Documenting the Cost and Price Analysis 145

ADEQUATE DOCUMENTATION 145
DOCUMENTING PRICE ANALYSIS 146
DOCUMENTING COST ANALYSIS 146
TYPICAL PRENEGOTIATION DOCUMENTATION 150
SUMMARY 150

Chapter 10 Negotiating the Transaction 153

INTRODUCTION 153
SITUATIONS WARRANTING NEGOTIATION 153
CONDUCTING EFFECTIVE AND EFFICIENT NEGOTIATIONS 154
MAINTAINING FAIRNESS AND EQUITY AMONG SUPPLIERS 155
NEGOTIATING CHANGE ORDERS 156
SUMMARY 158

Chapter 11 Documenting the Negotiation 161

NEGOTIATION MEMORANDUM 161
SUMMARY 163

Chapter 12 "Strategic Cost Analysis" Techniques Available to the Purchasing Manager 165

DEFINITION 165
CIRCUMSTANCES WHERE STRATEGIC COST ANALYSIS IS APPROPRIATE 165
SUMMARY 167

Glossary 169

Reference List 173

Appendix A Tables of Learning Curve Data   177

Appendix B Tables of Construction Job Factors   199

**Appendix C Sample Format for Postnegotiation
              Summary   203**

Index   205

Foreword

Organizations in all sectors of the modern economy are being pressured to decrease their operating costs while maintaining performance and product quality. In the private sector, both industrial and service businesses must keep costs as low as possible to remain competitive. In the public sector, government agencies must offer public services at a cost that neither raises taxes beyond what the public will bear nor raises the deficit. For both groups, the challenge is to keep quality up and costs down. Organizations that fail in this quest are likely to fail.

Sole-source providers, or suppliers in industries with high entry barriers, may have less pressure to maximize product or service quality for a given cost of operation. However, even customers held virtually captive by a sole source will eventually find a substitute product or service if dissatisfied. Newly established competitors have even succeeded against an entrenched government monopoly despite high capital entry costs because customers believed that the entrenched organization was charging too much for the service it delivered. This is the fate that awaits any organization that does not put sufficient effort into improving the cost/benefit ratio its customers realize from doing business with it.

There are many ways to cut costs. Some of them also involve cutting product quality, performance, or delivery, which is always dangerous. One such method is to decrease the quality of materials used to make the product or the skill of labor used to perform the service. Another is to lay off personnel until those remaining are so overworked or have such poor morale that their performance suffers. Yet another is to pressure a high (or at least adequate) quality supplier to cut prices so far that the supplier is unable by any means to break even. A supplier put in this position by a sufficiently powerful customer can be driven out of business.

The best ways for an organization to cut its costs involve careful analysis of what its true needs should cost, assuming reasonable economy and efficiency. In other words, the organization must decide the following:

1. what it really needs to do to remain in business,
2. what it really needs to have to do those things, and
3. what is a favorable but not unreasonably/impossibly low price for the inputs it needs.

These thought processes may be applied in either a tactical or a strategic manner. The tactical approach would naturally be short term in focus while the strategic approach would focus on the long term.

This book is a how-to manual covering specific procedures used to analyze organizational costs with a view to minimizing the cost of delivering a competitive level of quality. Most of the material herein will deal with the methodologies involved in tactical analysis—analysis designed to achieve the lowest reasonable cost on a single purchase. Strategic analysis, which ranges over the entire organization and involves all costs of doing business and is concerned primarily with institutionalized procedures and methodologies, is covered in chapter 12.

Deidre Maples

1
Introduction

PURPOSE OF THIS BOOK

This book is dedicated to providing advice to public- and private-sector buyers of both goods and services concerning those procedures and methods that will help them improve their effectiveness in obtaining fair and reasonable prices on their purchases. Specifically, it will address using a "Pricing Team"; methods of evaluating the market and bids/offers; price analytical/comparison techniques; various cost analytical methods; procedures for establishing realistic prenegotiation profit positions; using price, cost, and profit analysis in negotiations; and employing "Strategic Cost Analysis" to effect significant organizationwide cost savings.

THE BUYER'S PRICING TEAM AND THE RESPECTIVE ROLES AND RESPONSIBILITIES OF TEAM MEMBERS

Individual Buyers

Buyers are the most important members of the Pricing Team. Buyers are responsible for determining that the prices for their purchases are fair and reasonable (this is accomplished through the analyses that are the topic of this book), obtaining appropriate information for performing price and/or cost analysis from the prospective supplier, performing some form of price analysis on each procurement, performing a cost analysis on procurements in which the fairness and reasonableness of the price cannot be justified by price analysis alone, and negotiating the price if appropriate to the purchase.

Other Pricing Team Members

Some buyers, particularly those within the public sector and those working for large commercial or industrial firms, may have the benefit of assistance from several specialized members of the Pricing Team. Performing cost analyses, for example, normally requires the services of people specially trained to review cost proposals. These individuals are generally trained in accounting and auditing and may indeed be auditors. Some large organizations train people with business, accounting, and finance backgrounds to prepare estimates and perform cost/price analysis. They are often called cost/price analysts or cost estimators. Auditors, cost/price analysts, and estimators are generally responsible for reviewing and evaluating the cost elements and proposed profit/fee of a supplier's proposal when requested by the buyer. The person assigned this task should request as much assistance from engineering or other technical departments as needed to properly evaluate manhours and materials proposed by the supplier; conduct an audit or review the supplier's proposed labor rates, burdens, overheads, and other cost elements; and/or request an audit of the supplier's proposal as required. The auditor–analyst–estimator is also generally charged with reviewing Price–Negotiation Memoranda on procurements of significant dollar value to assure that all cost–financial guidelines in the buying organization's policies are followed. Specialized purchases may require expertise in packaging, transportation, quality assurance, or other disciplines.

DEFINITIONS

A comprehensive list of the basic definitions that will be used throughout this book can be found in the glossary of terms. The four most important terms, which we shall explore at some depth, are price analysis, cost analysis (with its accompanying profit analysis), strategic cost analysis, and transactional analysis. These definitions are restated here for your benefit:

Price analysis is the process of examining and evaluating a proposed price without evaluating its separate elements of cost and profit. It may be accomplished by the following comparisons, which are listed in their relative order of preference:

- Comparison with other prices and quotations submitted.
- Comparison with published catalog or market prices.

- Comparison with prices set by law or regulation.
- Comparison with prices for the same or similar items.
- Comparison with prior quotations for the same or similar items.
- Comparison with market data (indexes).
- Application of rough yardsticks (such as dollars per pound or per horsepower or other units) to highlight significant inconsistencies that warrant additional pricing inquiry.
- Comparison with independent estimates of cost developed by knowledgeable personnel within the buying organization.
- Use of value analysis.
- Use of visual analysis.

Cost analysis is the process of reviewing and evaluating the separate cost elements included in an offeror's proposal, including the judgmental factors applied by the offeror in projecting from historical cost data to the estimated costs included in the proposal. Cost analysis is done to help the buyer form an educated opinion on the degree to which the proposed costs represent what the contract should cost, assuming reasonable economy and efficiency. It includes verifying cost data and evaluating cost elements, including:

- The necessity for and reasonableness of proposed costs.
- The offeror's projection of cost trends on the basis of current and historical cost or pricing data.
- A technical appraisal of the estimated labor, material, tooling, and facilities requirements and of the reasonableness of scrap and spoilage factors.
- The application of approved indirect cost rates, labor rates, or other factors.

Among the evaluations that should be made, where the necessary data are available, are comparisons of an offeror's current estimated costs with:

- Actual costs previously incurred by the same supplier or offeror.
- Previous cost estimates from the offeror or from other offerors for the same or similar items.
- Other cost estimates received in response to the solicitation.
- Independent cost estimates by technical personnel.
- Forecasts of planned expenditures.

Strategic cost analysis is characterized as the sum total of the broad-based, organizationwide plans and analyses designed to increase purchasing savings in the long run. This type of analysis is not related to specific purchasing transactions.

Transactional analysis is the narrowly based plans and analyses performed

4   Cost/Price Analysis: Tools to Improve Profit Margins

by the individual buyer (with expert assistance, as appropriate) to increase purchasing savings in the instant purchase order or contract.

Although this book will concentrate on the analysis of instant purchase transactions (transactional analysis), a chapter will be devoted to the more broad-based cost analytical approach, which we define here as strategic cost analysis.

COMPARISON OF DIFFERENT ANALYTICAL METHODS FROM THE VIEWPOINT OF COMPLEXITY AND QUANTITY OF RESOURCES REQUIRED

Because the author is a firm advocate of the adage "A picture is worth a thousand words," a graphic representation of these terms is presented in Figure 1-1. The diagram will also be used as a skeletal outline for the discussions that follow. The diagram presents the different types of analysis

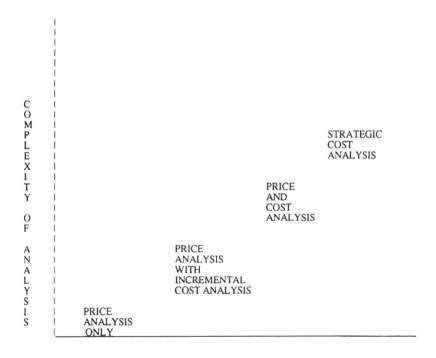

FIGURE 1-1.

that will be discussed in this manual. All but strategic cost analysis are considered to be methods of transactional analysis.

Price analysis is generally the simplest transactional analysis technique and is considered the least resource intensive. Price analysis can be conducted in a relatively short period of time and can normally be conducted by the buyer without any outside assistance from members of the Pricing Team. Price analysis with incremental cost analysis adds complexity to the analysis and requires more resources by virtue of the fact that certain elements of cost (but not necessarily all) require some degree of cost analysis. This technique is commonly used when the buyer is purchasing modified commercial items. The basic unmodified item can be analyzed by using price analysis and the modifications can be analyzed by partial cost analysis. The buyer can often perform this type of analysis on his or her own, but may require assistance from one or more members of the Pricing Team. Price and (full-blown) cost analysis is still more complex, usually invoking other members of the Pricing Team. This type of analysis is appropriate in sole- and single-source purchases of significant magnitude and complexity. Cost-analytical efforts, and the negotiations that invariably ensue, may extend for several weeks or even months. Strategic cost analysis is the most complex and resource intensive of the four analytical methods. This type of analysis takes the long-haul approach not keyed to specific purchase transactions. Buyers undertaking strategic analysis will be dealing with fundamental business systems and procedures, generally of a type that must be institutionalized before they can be effective. A formalized make-or-buy system used by the organization to assure cost-efficient sourcing of materials and services would be one example of such a system. Additional examples are discussed in Chapter 12.

TYPES OF MARKETS AND SUPPLIER PRICING STRATEGIES AND THE ANALYTICAL METHODS APPROPRIATE TO EACH

Preparing Buyer In-House Estimates

Although different organizations have different policies concerning the degree and type of documentation needed to support a pricing decision, most organizations, public and private, insist that a cost estimate be prepared by requesters on every purchase as an aid to pricing. This estimate is normally a detailed, bottom-up estimate for services (architect engineering, construction, and so forth), and for specially engineered or fabricated hardware or one-of-a-kind items. In most supply and material purchasing, an estimate based on previous prices paid often suffices. Preparation of an in-house estimate by the buying organization is a good mental discipline, requiring the buying organization to place itself in the supplier's shoes. Understanding the supplier's competitive situation is important to determine the appropriate analytical method. There is a uniform policy within the purchasing community that every purchase transaction, regardless of dollar amount, should have a determination that the price is fair and reasonable. This determination is based on price analysis (often using the in-house estimate as a basis for comparison) supplemented by cost analysis if necessary. Documentation of the facts that support a price reasonableness determination (to include a copy of the in-house estimate) should be included in the purchase file.

Degree of Competition in Markets

Most goods and services are bought in a competitive marketplace where the forces of competition cause the supplier to price goods and services according to the competition. In these situations, the supplier is less interested in costs and profit margin than in the prices being charged by the competition. Effective competition is most pronounced in those industries having few entry barriers. Even the American automobile industry provides effective competition in what is commonly referred to as an "oligopolistic industry" (few sellers and many buyers). Where the buyer can rely on competition to set the price, price analysis alone will suffice to assure reasonableness of price. A cost analysis may be required to supplement the price analysis, particularly in situations where price competition among offerors is nonexistent or questionable. Such a situation is often prevalent in purchases of highly specialized or extremely technical equipment or for certain services available from limited sources (specialized consulting services come readily to mind). Because cost analysis is such a costly and administratively burdensome process, it should generally not be employed when reasonableness of price can be established by adequate price competition or (as an acceptable substitute) by comparison with catalog or market prices of commercial items sold to the public in substantial quantities. Another special situation is purchasing regulated utility services. Price competition is not present, and cost analysis is unnecessary because of the involvement of public utility commissions and other regulatory bodies in the rate-setting process. Regulated utilities provide services at prices set by law or regulation. Price analysis alone (comparing the offered prices against the regulated rate) is sufficient in those instances.

Matching Buyer Analytical Method With Supplier Pricing Method

The buyer must understand the degree of competition in the marketplace to determine the method used by the supplier to price his or her products or services. In a market where there is effective competition, the supplier will generally price to the competition with less regard for cost and profit. The forces of competition will operate to keep the supplier's offered prices fair and reasonable. Because the supplier bases his/her price on a bottom-line basis, the buyer should respond by using a bottom-line analytical method: price analysis. If, however, the supplier is free to pass on to the buyer all costs plus a substantial profit as a result of operating in a market without effective competition, the buyer should respond to this bottom-up pricing with a bottom-up analytical method: cost and profit analysis, supplemented with price analysis. This concept is illustrated in Table 1-1.

Notice the one analytical method appropriate to all markets is price analysis. Where there is ineffective competition, price analysis should be used to supplement cost analysis because price analysis answers the question: "What would the supplier charge if he or she were operating in a competitive marketplace?" Many experienced analysts maintain that price analysis is more powerful than cost analysis for that very reason. They also believe it is possible to generate multiple negotiating positions more easily using both cost and price analysis than it is using cost analysis alone.

SUMMARY

The buyer is ultimately responsible for determining that the prices he/she pays are fair and reasonable. This is accomplished through price analysis and/or cost analysis of the individual purchase transaction. The degree and extent of the analysis performed depends on the circumstances. In certain circumstances, particularly those where extensive cost analysis is required, the buyer will be assisted by specialized functional personnel, to include auditors, cost/price analysts, estimators, engineers, and/or other technically oriented personnel. The results of the price and/or cost analysis of the individual purchase transaction will be used by the buyer to negotiate price and other terms with supplier(s).

TABLE 1-1

If the Supplier Estimates Based on:	The Buyer Should Use:
Bottom-line competitive pricing	Bottom-line price analysis
Bottom-up cost & profit buildup	Bottom-up cost and profit analysis (with price analysis)

In addition to transactional analysis, which concentrates on price and cost analysis of individual purchase transactions, the buyer and purchasing management will be challenged to effect major, organizationwide cost savings through the use of strategic cost analysis. This type of analysis takes the long-haul approach not keyed to specific purchase transactions. Buyers undertaking strategic analysis will be dealing with fundamental business systems and procedures, generally of a type that must be institutionalized before they can be effective. Various strategic cost analytical methods, including a formalized make-or-buy decision system, are discussed later in the book.

Preparing an in-house estimate requires the buying organization to place itself in the supplier's shoes. Understanding the supplier's competitive situation is important in determining the appropriate analytical method. Every purchase transaction, regardless of dollar amount, must have a determination that the price is fair and reasonable. This determination is based on price analysis (often using the in-house estimate as a basis for comparison) supplemented by cost analysis if necessary.

Most goods and services are bought in a competitive marketplace where the forces of competition cause the supplier to price goods and services according to the competition. In those situations where the buyer can rely on competition to set the price, price analysis alone will suffice to assure reasonableness of price. Although price analysis generally suffices to document price reasonableness, a cost analysis may be needed to supplement price analysis, particularly when the presence of price competition among offerors is nonexistent or questionable. Such a situation is often prevalent when purchasing highly specialized or extremely technical equipment or for certain services available from limited sources (specialized consulting services come readily to mind).

The buyer must understanding the degree of competition in the marketplace in order to determine the method used by the supplier to price products and/or services. If the supplier is operating in a market where there is effective competition, the supplier will generally price to the competition, and the buyer should respond by using price analysis. If, however, the supplier is free to pass on to the buyer all costs plus a substantial profit as a result of operating in a market without effective competition, the buyer should respond by using cost and profit analysis supplemented with price analysis.

The one analytical method appropriate to all markets is price analysis. Where there is ineffective competition, price analysis should be used to supplement cost analysis because price analysis serves to answer the question: "What would the supplier charge if he or she were operating in a competitive marketplace?" Price analysis is also helpful in generating multiple negotiating positions.

2
Types of Contracts and the Analytical Methods Appropriate to Each

INTRODUCTION

Negotiating price is meaningless unless that price is related to the type of contract to be awarded. A price that appears fair and reasonable on a firm-fixed-price basis may be unreasonable on a cost-reimbursement contract (which generally has a lower degree of risk for the supplier). Like price, contract type is generally negotiable. This fact often requires the buyer to prepare a different set of price objectives for each of the contract types considered in negotiation.

Another pricing factor impacting the type of contract selected is the matter of prospective versus retroactive pricing. Firm-fixed-price contracts lend themselves readily to firm pricing in advance of award. Contracts other than firm-fixed-price (sometimes called flexibly priced contracts, to include incentive and cost-reimbursement contracts) generally contemplate not only prospective pricing (to establish target cost estimates, fee amounts and ceilings), but also retroactive pricing (after award and after performance/delivery) for certain purposes. Cost-reimbursement contracts require the buyer (aided by his or her Pricing Team) to review periodic invoices or vouchers to screen for unallowable costs actually incurred and billed. These contracts also require some form of close-out audit after the contract is complete to finalize indirect cost rates and do a final check for unallowables. Incentive contracts are generally structured to permit invoicing on a fixed-price basis (without vouchers). Just as for cost-reimbursement contracts, a close-out audit is performed at contract completion to finalize indirect cost rates and screen for unallowables. Buyers must plan for heavy postaward involvement in pricing for certain types of contracts. Heavy postaward pricing involvement makes these types of contracts administratively burdensome.

10 Cost/Price Analysis: Tools to Improve Profit Margins

THE EFFECT OF RISK ON SELECTING CONTRACT TYPE

Assessing risk is critical in selecting the type of contract. Buyers must get project office technical support in doing risk analysis on complex procurements. They will jointly participate in preparing input for the buyers' management on these types of procurements. The objective is to select a type of contract with the greatest incentive for efficient and economical performance. When there is significant uncertainty in the purchase transaction, both the buyer and the supplier will try to pass as much of the cost and technical risk as possible on to the other contracting party. In such situations, a firm-fixed-price will generally not satisfy the needs of both parties to minimize their exposure and risk. Different types of contracts represent varying degrees of risk to the buyer and supplier. Figure 2-1 attempts to portray this concept for four commonly used contract types.

OTHER FACTORS INFLUENCING SELECTION OF CONTRACT TYPE

Supplier Motivation

Some companies are so accustomed to working under cost-reimbursement contract arrangements they develop a mind-set that prevents them from developing the mentality required to work under fixed-price type contracts. Other companies are just the opposite. They are accustomed to dealing at an arm's length with their buyers on a firm-fixed-price basis and are unwilling

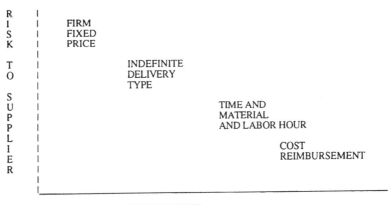

FIGURE 2-1.

to accept the buyer intrusion that invariably accompanies the award of a cost-reimbursement contract. Supplier bias must be kept in mind in selecting an appropriate type of contract.

Competitive Environment

Although there are situations where cost proposals are necessary to support a firm-fixed-price in the face of ineffective competition, it is more common to award a firm-fixed-price contract as a result of effective competition. As a general rule, firm-fixed-price contracts (and their companion fixed-price with economic price adjustment contracts) are awarded after price analysis (only) has been used to assure fairness and reasonableness of price. A corollary of this is that contracts of other types will be awarded in the face of ineffective competition and hence will require some degree of cost analysis. Although a buyer should not avoid using other than firm-fixed-price contracting, he or she should justify explicitly (in the negotiation memorandum or elsewhere) the choice of a contract type other than firm-fixed-price.

Supplier's Accounting System

Flexibly priced contracts (other than firm-fixed-price) and cost-reimbursement contracts all carry with them some degree of retroactive pricing. As such, they rely on a fair and accurate allocation of cost to the work required by these types of contracts. Suppliers whose cost accounting systems cannot accurately allocate costs to jobs should generally not be considered for these type contracts.

Procurement Method

Most buyers are trained to use competitive bidding, with the source-selection decision based on lowest price. The resulting contract type is generally firm-fixed-price. Other contract types, particularly cost-reimbursement contracts, should be considered when competition is either absent or very limited and other factors than price are being considered in the selection process. Although it is not impossible, it would be extremely difficult to award a cost-reimbursement or other flexibly priced contract as a result of competitive bidding.

Administrative Workload

Few buying organizations plan adequately for the massive amount of administration required by many different contract types, particularly cost-reimbursement contracts. Cost-reimbursement contracts are often for complex products or services involving expenditures of several hundreds of thousands (or even millions) of dollars. They are commonly thought to require a true buyer-supplier partnership. Cost-reimbursement contracts often require five to ten times the number of contract administration personnel and level of resources that a firm-fixed-price contract would need.

FIXED-PRICE FAMILY OF CONTRACTS

Firm-Fixed-Price

A firm-fixed-price contract provides for a price that is not subject to any adjustment on the basis of the supplier's cost experience in performing the contract. This contract type places on the supplier full risk and responsibility for all costs and resulting profit or loss. It provides maximum incentive for the supplier to control costs and perform effectively, and imposes a minimum administrative burden on the buyer. It is generally suitable for acquiring commercial or commercial-type products, or for acquiring other supplies or services on the basis of reasonably definitive functional or detailed specifications when fair and reasonable prices can be established at the outset.

A typical schedule for a firm-fixed-price contract is shown in Table 2-1.

Fixed-Price with Economic Price Adjustment/Escalation

A fixed-price contract with economic price adjustment provides for upward and downward revision of the stated contract price upon the occurrence of

TABLE 2-1 Schedule of Firm-Fixed-Price Services

Schedule

Perform construction services according to the specification herein. The completed facility shall be completed and turned over to the buyer NLT 31 December 1994. The Supplier's work under the firm-fixed-price line item of the contract shall be considered complete when the completed facility has been turned over complete without deficiencies of any kind. (See Paragraph 1.2.1, Specification).

Item No.	Supplies/Services	Quantity	Unit	Unit Price	Amount
1	Warehouse Built According to Attached Specification	1	Ea	$25,000	$25,000

Types of Contracts and the Analytical Methods Appropriate to Each 13

specified contingencies. Economic price adjustments are of three general types:

- *Adjustments based on established prices.* These price adjustments are based on increases or decreases from an agreed-upon level in published or otherwise established prices of specific items or the contract end items.
- *Adjustments based on actual costs of labor or material.* These price adjustments are based on increases or decreases in specified costs of labor or material that the supplier actually experiences during contract performance.
- *Adjustments based on cost indexes of labor or material.* These price adjustments are based on increases or decreases in labor or material cost standards or indexes that are specifically identified in the contract.

A fixed-price contract with economic price adjustment may be used when (1) there is serious doubt concerning the stability of market or labor conditions that will exist during an extended period of contract performance; and (2) contingencies that would otherwise be included in the contract price can be identified and covered separately in the contract. Price adjustments based on established prices should normally be restricted to industrywide contingencies. Price adjustments based on labor and material costs should be limited to contingencies beyond the supplier's control. A fixed-price contract with economic price adjustment should not be used unless it is necessary either to protect the supplier and the buyer against significant fluctuations in labor or material costs or to provide for contract price adjustment in the event of changes in the supplier's established prices.

Fixed-Price Incentive

A fixed-price incentive contract is a fixed-price instrument that provides for adjusting profit and establishing the final contract price by applying a formula based on cost as incurred and normally on performance factors as well. The final price is subject to a price ceiling negotiated at the outset. A fixed-price incentive contract is appropriate when a firm-fixed-price contract is not suitable and the nature of the supplies or services being acquired and other circumstances of the purchase are such that the supplier's assumption of a degree of cost responsibility will provide a positive profit incentive for effective cost control and performance. If the contract also includes incentives on technical performance and delivery, the performance requirements must provide a reasonable opportunity for the incentives to have a meaningful impact on the supplier's management of the work. A fixed-price incentive contract should be used only when the following occur:

14   Cost/Price Analysis: Tools to Improve Profit Margins

- This contract type is likely to be less costly than any other type.
- It is impractical to obtain supplies or services of the kind or quality required without the use of this contract type.
- The supplier's accounting system is adequate for providing data to support negotiating final cost and incentive price revisions.
- Adequate cost information for establishing reasonable firm targets is available when the contract is initially negotiated.

"The Wayne Corporation" case study illustrates some of the more important points about fixed-price-incentive contracts. The case study is continued after the discussion of cost-incentive contracts (see under cost-reimbursement family of contracts).

The Wayne Corporation

Power, Inc. has developed a requirement to produce a quantity of 20 lightweight, high-power output, power supplies. The power supply has been fully developed and qualification tests conducted. Of note is the fact that although the item eventually passed rigid test requirements, retest of each article was required due to the system's insufficient cooling capability. (After individual engineering efforts the system turned out reliable and capable of effective use.) The buyer, after consultation with the requisitioner, had made the determination that a fixed-price-incentive contract type was appropriate. Analysis of cost data has led to the following objective for negotiating with the sole source supplier, The Wayne Corporation:

Target cost $209,000
Target profit   11,000
Target price   220,000

Based on an analysis of uncertainty, it appeared reasonable that the chance of an overrun of the target cost was as great as the chance of an underrun; and that a point of total assumption should be established at $224,000. At this point, they felt The Wayne Corporation should receive $5,000 in profit.

Given this information, let us determine the following:

- The share ratio.
- The ceiling price.

Discussion

The share ratio reflects the degree to which the buyer and the supplier will be responsible for cost in excess of the target cost and the degree to which the buyer and supplier will share in cost savings if the target cost is underrun.

Types of Contracts and the Analytical Methods Appropriate to Each 15

Although it is possible to have separate share ratios above and below the target cost, it is more common to use one ratio throughout the potential cost range. The share ratio is normally expressed by writing the buyer's share in the numerator and the supplier's share in the denominator. The % is normally omitted.

A typical share ratio would be 70/30, which means the buyer pays 70 percent of every dollar overrun and gets the benefit of 70 percent of every dollar underrun while the supplier pays 30 percent of every dollar overrun and gets the benefit of 30 percent of every dollar underrun. The supplier's share ratio is normally mathematically determined by graphically plotting the data. Cost is normally plotted on the X or horizontal axis and profit is normally plotted on the Y or vertical axis. Two data points are necessary to compute the slope. In our example, we have one data point at $209,000 cost and $11,000 profit and another data point at $224,000 cost and $5,000 profit. The slope is computed by taking the difference in profit between the two points and dividing by the difference in cost between the same two points. The change in profit is $11,000 − $5,000, or $6,000. The change in cost is $224,000 − $209,000, or $15,000. The resulting supplier share percentage or ratio is $6,000/$15,000 or 40 percent. The buyer's share is computed by subtracting the supplier's share from 100 percent. In our case, the result is 100 percent − 40 percent or 60 percent. The share formula in our case is thus "60/40" (buyer share divided by supplier share).

The ceiling price in a fixed-price-incentive contract is the maximum price to be paid by the buyer. It reflects a break-even point for the supplier, since the profit at that point is 0. At the point of total assumption ($224,000 in this example), the supplier starts a 0/100 share: every dollar spent beyond $224,000 is paid by the supplier. The supplier can overrun this point by $5,000 (the profit at that point) before losing money. The ceiling price is easily computed, by taking the value of the point of total assumption and adding the profit at that point. The result in our example is $224,000 + $5,000, or $229,000.

The graphic representation of this fixed-price-incentive contract is shown in Figure 2-2.

Let's change the previous example to make the ceiling price $220,000, the share ratio 70/30, the target cost $200,000, and the target profit $10,000.

Given this information, let us determine the point of total assumption.

Discussion
The point of total assumption can be computed mathematically by the following formula:

$$\text{Point of Total Assumption} = \frac{(\text{Ceiling Price} - \text{Target Price})}{(\text{Buyer Share})} + \text{Target Cost}$$

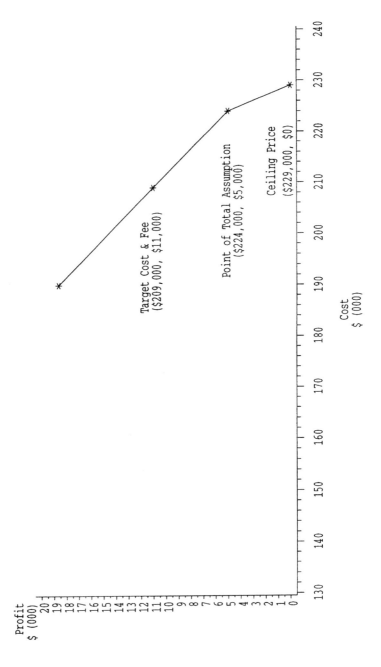

FIGURE 2-2.

(Remember Target Price is Target Cost Plus Target Profit).

In our example, this would be: $\dfrac{(\$220{,}000 - \$210{,}000)}{(.70)} + \$200{,}000$

Or: $\dfrac{\$10{,}000}{.7} + \$200{,}000$. The result is $14,285.71 + $200,000 or $214,285.71.

The point of total assumption could have been graphically determined by plotting the target cost/target profit point and the ceiling price (on the X axis, since the profit at this point is 0) and then by drawing the share line through the target cost/target profit point at a 30 degree angle (the slope of the share line is the supplier's share percentage—30 percent in this case). This line would reflect a loss of 300 dollars profit for every 1,000 dollars increase in cost past the target cost. After the share line is drawn, the $0/100$ share line should be drawn upward from the ceiling price point on the X axis. The intersection of these two lines is the graphic representation of the point of total assumption.

The graphic representation of the data for this slightly different fixed-price-incentive contract is shown in Figure 2-3.

Indefinite-Delivery (Blanket Order) Type

Indefinite-delivery (blanket order) type contracts are often called "open-end" or "term" contracts. They are quite frequently employed in service contracting, either in pure form or in combination with other types such as firm-fixed-price and time-and-materials or labor-hour. There are three basic forms of indefinite-delivery (blanket order) type contracts: indefinite-quantity/indefinite-delivery, requirements, and definite-quantity/indefinite-delivery. Since the first two are more commonly used, let's explore these two in some detail.

Indefinite-Quantity/Indefinite-Delivery (IQID)

The IQID establishes firm-fixed-unit prices for the units of work sought by the buyer. It states a guaranteed minimum quantity (base amount) and an estimated maximum quantity (ceiling or cap). This contract type is used when the buyer is seeking a service that requires high mobilization or start-up costs, which would not be recoverable if only a small percentage of the total estimated services were actually ordered. The guaranteed minimum quantity should be more than just a nominal quantity, but should not exceed the total amount the buyer feels certain will be needed. This base provides the service supplier with a minimum upon which to offer, thus providing a means to recover the costs of mobilization or start-up.

When an IQID contract is awarded, the buyer is generally authorized to

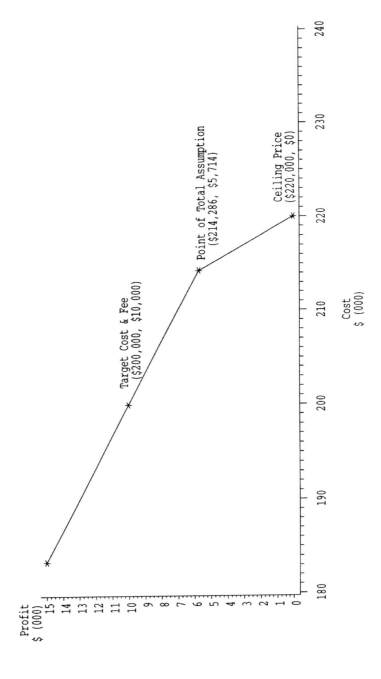

FIGURE 2-3.

order any number of units up to the maximum amount estimated for each work item in the schedule. The supplier is then responsible for supplying the guaranteed minimum quantity and additional quantities up to that maximum. If the buyer fails to order the guaranteed minimum quanity, the supplier must still be paid the value of the guaranteed minimum quantity.

An IQID contract provides the buyer with certain benefits. If the supplier's performance is poor, the supplier may be paid for the guaranteed minimum quantity and the contract closed. A new solicitation and subsequent contract for the service can then be issued. The buyer should be aware, however, that in a new solicitation, there is no guarantee that the same supplier will not be the lowest offeror.

Funds for this type of contract are obligated only for the amount guaranteed at the time of award. The balance of the funds for the units of work in excess of the base are obligated through delivery orders as the items or services are procured by the buyer.

A typical schedule for an indefinite-quantity/indefinite-delivery contract shown in Table 2-2.

TABLE 2-2   Schedule of Indefinite-Quantity/Indefinite-Delivery

Schedule					
0002	Indefinite-Quantity Type Work or Services. Price for labor and material to perform work or services in the Base Period to perform the Unit Priced Tasks listed below as specified in the Statement of Work, paragraphs 1.6.1.1 through 1.6.1.10. The quantities listed below are realistic estimates provided solely for the purpose of offer evaluation. The price for this bid item is the total of the subline items listed in the Schedule of Indefinite Quantity. The guaranteed minimum quantity (which will be ordered during the Base Period) shall be computed in dollars at 20 percent of the estimated total price for contract line item 0002. There will be no minimum guarantee quantity for individual subline items. The clause in this contract entitled "Indefinite Quantity" explains this type of contract and its operation.				
Item No.	Supplies/Services	Estimated Quantity	Unit	Unit Price	Amount
0002A	Provide order-filling services (per para.1.6.1.1 through 1.6.1.3)	1000	40×60 in. Charts	$ __	$ __
0002B	Provide order-filling services (per para.1.6.1.4 through 1.6.1.7)	1000	20×30 in. Charts	$ __	$ __
0002C	Provide order-filling services (per para. 1.6.1.8 through 1.6.1.9)	1000	Binder Size charts	$ __	$ __
0002D	Provide order-filling services (per para.1.6.1.10)	1000	Computer Diskettes	$ __	$ __
	Total price for contract line Item 0002 (0002A Through 0002D)				$ __

Let us investigate a real-world situation calling for the application of this type of contract. Using historical data from maintenance records, a buyer determines that between 200 and 300 vehicles will require repair during the year. The number of repairs varies based on the amount of work that takes place. The buyer knows the approximate amount of work required based on historical data, that the nature of the work to be done can be defined, and that sound estimates for maintenance can be established prior to contracting. The buyer does not know when the work will be required during the year, what specific types of maintenance will be required, or what the exact number of maintenance visits will be. An IQID contract type is appropriate because this situation calls for a fixed-unit-price contract where the unit price is not subject to change. In this example, neither the quantity of work nor its frequency is definite. In addition, a variety in types of maintenance service are involved and it would be in the buyer's best interest to obtain a separate price for each type of maintenance service if possible. The buyer should contact the Facilities Manager and obtain a more specific breakdown of the types of maintenance service along with a historical estimate of the quantities involved. In this way, the offeror will have some insight into the amount of additional equipment required (if any). The indefinite quantity contract would provide the buyer with a means of guaranteeing to the supplier that the mobilization costs and some portion of the equipment costs could be realized regardless whether the need actually equaled the estimate. The contract also provides the buyer with a means of ordering the service on an as-needed basis. The risk to the buyer associated with this type contract is greater than that under the firm-fixed-price type because of the need to issue delivery orders for all work exceeding the guaranteed minimum quantity.

Requirements
The requirements contract establishes firm-fixed-unit prices for the units of work sought by the buyer. This type contract provides that quantities stated in the schedule are estimated and are not purchased by the contract award. This type contract further provides that the stated quantities are used only for the purpose of evaluating offers and determining the low offeror. Offerors are put on notice that all or none of the work may be ordered. This type contract also provides a guarantee that if any contract work listed in the schedule is needed during the term of the contract, such will be procured from this supplier. This procurement guarantee excludes performance of such work by the buyer in-house, unless the contract states that under specified circumstances the buyer reserves the right to use in-house employees. Such reservations must be very explicit.

The solicitation package for a requirements contract must clearly state that estimated quantities shown are solely for offering and offer evaluation purposes. The solicitation documents include a reasonable maximum ceiling or cap that may be ordered overall and limits both the amount of any delivery order and the number of delivery orders at any one time under the contract. This cap or ceiling protects the supplier from being inundated by an unanticipated work load and allows the buyer to solicit separately for large jobs. The requirements contract does not obligate funds at the time of contract award. Obligation of funds occurs upon issuance of a delivery order. When the item or service is required, a delivery order is written pursuant to the terms of the contract. The requirements type contract does not provide the supplier with any insight of how to gear up for the work; it only provides an estimate on which to base minimum levels of personnel, supplies, and equipment. As a result of the unknowns associated with this type of contracting, its use may attract few offerors.

A typical schedule for a requirements contract is shown in Table 2-3.

Let us investigate a real-world situation calling for the application of this type of contract. A buyer is required to maintain 100 acres of grassed area. Historical data indicates that during the growing season there may be extended periods of time in which no rain falls during some years. It is estimated that during a normal rainfall year the grass must be cut at least 28 times, but experience shows dry years in which it needed to be cut only 14 times. The

TABLE 2-3   Schedule of Requirements

Schedule					
0001	Requirements-Type Work or Services Price for labor and material to perform work or services in the Base Period to perform the Unit Priced Tasks listed below as specified in the Statement of Work, paragraph 1.3.1 through 1.3.3. The quantities listed blow are realistic estimates provided solely for the purpose of offer evaluation. The price for this bid item is the total of the subline items listed in the Schedule of Requirements. There is no minimum guarantee for this line item. The clause in this contract entitled "Requirements" explains this type of contract and its operation.				
Item No.	Supplies/Services	Estimated Quantity	Unit	Unit Price	Amount
0001A	Provide exterior painting services (per para.1.3.1)	1,000	SY$ __	$ __	$ __
0001B	Provide roof repair services (per para.1.3.2)	2,000	SF$ __	$ __	$ __
0001C	Replace damaged pipes (per para. 1.3.3)	1,000	LF$ __	$ __	$ __
	Total price for contract line Item 0001 (0001A THROUGH 0001C)				$ __

frequency of the cuts depends on the rainfall. Three days of heavy rain and a day of sunshine makes the grass grow so rapidly it may require cutting on the fifth day rather than on the seventh day. If a dry spell follows, the grass may not need to be cut for ten or more days. The buyer would like to control the frequency by ordering the cuttings only when they are needed. The buyer knows that the task of grass cutting is easily definable, that a reasonable estimated quantity can be determined, that 100 acres require the service (note that the frequency of the cuts is the variable in this problem—not the acres to be cut), and that a sound estimate of the cost per cut can be established prior to contracting. The buyer does not know the accuracy of the estimate because of the weather variance or at what intervals the work will be required. A requirements contract is appropriate here because this is a fixed-unit-price contract situation where the unit price is not subject to change. Since no obligation of funds occurs until a delivery order is issued, the actual amount of work is fully controlled by the buyer as provided for under the terms of the contract. The risk to the buyer under this type of contract is greater than under the firm-fixed-price type because of the need to issue delivery orders for all work.

Estimating Work Units for Indefinite-Delivery (Blanket Order) Type Contracts

Indefinite-delivery (blanket order) type contracts require an estimate of the work units the buyer will probably procure during the contract period. The buyer, based on funding constraints associated with the project, estimates using available historical data. The importance of basing this estimate on historical data cannot be overemphasized. Inaccurate estimates increase the risk for the buyer. The supplier needs to build into the unit price the direct costs associated with delivering the items or service and a reasonable margin of profit. The unit price developed by the offeror is based on mobilization costs, supply and equipment costs, overhead costs, labor costs, and a profit. In many cases, mobilization and equipment costs are constant regardless of the number of units required. Overestimates of the quantity result in low unit prices and underestimates result in high unit prices.

When an underestimate is made, the supplier's unit cost will be higher to cover the fixed costs distributed over the quantity of work. If the buyer exceeds the estimate, such excess work will prove more costly to the buyer. In this case, the inflated unit price will be high by comparison against similar procurements of larger quantities. Conversely, an overestimate of the buyer's needs will generate a lower unit price. When the buyer does not call for the units estimated, the supplier begins to suffer losses since mobilization costs, equipment costs, and profit cannot be realized. The result is a disgruntled supplier and a potential claim against the buyer. A buyer does not want to cause a service supplier hardships.

Ordering Considerations for Indefinite-Delivery (Blanket Order) Type Contracts

When considering the use of indefinite-delivery (blanket order) type contracts the following definitions must be taken into account and not confused with each other:

- Minimum order is the minimum quantity of any item under any one delivery order/authorization. The minimum should be set high enough to make accomplishment economically feasible for the supplier.
- Maximum order (including a maximum number of orders) is the maximum quantity of any item under any one delivery order/authorization, including the maximum number of work orders/authorizations to be in effect at any one time. This limitation should be set to prevent the supplier from becoming inundated with orders to the point the supplier cannot efficiently perform.
- Minimum guaranteed quantity is a guarantee, applicable only to indefinite-quantity/indefinite-delivery type contracts, that establishes the dollar value of the contract by stating the minimum dollar amount that the buyer will pay even if the total of all delivery orders/authorizations does not equal that amount. *Do not confuse with minimum order.*
- Maximum contract value is the ceiling or cap in dollars or quantity percentages as stated that the value of the contract cannot exceed. This cap should normally be in the realm of 200 percent over the estimated quantities being used as the basis for establishing the low offeror.

Delivery Orders for Indefinite-Delivery (Blanket Order) Type Contracts

In indefinite-delivery (blanket order) type contracts, the demand for the items/service is based on *need* rather than on a *prearranged* delivery schedule. For this reason, the supplier must be made aware in the contract document how the scheduled items/work are to be ordered. This description should be included in the solicitation package. Delivery orders (sometimes called task orders, calls, or releases) should be supported by a funding document (requisition) at time of issue. Delivery orders not only establish what items/work will be delivered by the supplier but also establish a financial obligation on the part of the buyer. The delivery order under an IQID contract does not obligate funds until the guaranteed minimum quantity is reached. After the minimum is reached, it obligates funds as it always does in requirements contracts. To complete the obligation/payment cycle, a copy of each delivery order will usually accompany the invoice when submitted to Accounts Payable to initiate payment. Two basic methods of ordering are available to the buyer.

- Order for supplies or services: Most purchase order forms provide the buyer with a means of furnishing the supplier with all the information related to the items/work being ordered, including accounting data chargeable.
- Letter format: The letter format must include specific statements of work or items to be ordered (normally right from the schedule), the quantities, the value of each line item of work (taken from the schedule) and the overall total, start and completion dates (for services) or delivery dates (for supplies), complete accounting data as chargeable to the various cost codes for the supplies or services ordered, the contract number under which the order is placed, and the delivery order number in consecutive sequence for each delivery order associated with the contract.

The proper method to effect a delivery order for service and not be trapped by internal auditors is to issue Delivery Order No. 1 in either form at the beginning of the first month of the contract term. The delivery order should set forth a dollar amount (about $\frac{1}{2}$ of estimated contract price or the amount then available if less than the estimated contract price) and state "For Work to be Authorized" during the month of _____ (fill in coming month as appropriate). After work is authorized (keep a record so the amount of Delivery Order No. 1 is not exceeded), completed, and accepted for that month and payment can be made, Delivery Order No. 1 (Modified) should be issued. The modified delivery order should itemize the work to be paid for, set forth the quantities, unit prices and totals therefore, and ultimately increase or decrease the amount of money established by Delivery Order No. 1.

Comparison of Different Indefinite-Delivery (Blanket Order) Type Contracts.

The following list compares IQID and requirements contracts with respect to their essential elements, limitations, and suitability.

1. Indefinite-Quantity
    a. Essential elements:
       Use where it is impossible to determine in advance the precise quantities of supplies or services that will be needed by designated activities during a definite contract period.
       A base or target amount of each item is included for offer evaluation purposes.

A stated minimum (guaranteed) shall be ordered by the buyer during the contract period. Further, a maximum amount to be ordered is to be specified.

Method of ordering work must be stated as well as minimum/maximum orders allowable during a specified period of time.

Minimum amount stated must be more than a nominal quantity (reasonable), yet it should not exceed the amount the buyer is fairly certain to order.

   b. Limitation:

Funds are obligated by minimum (guaranteed) amount and thereafter by individual orders.

Provides flexible quantity and delivery schedule with limited buyer obligation.

A fixed-unit-price schedule (Schedule of Prices), which provides basis of cost items to be ordered, is required prior to award.

Minimum amount stated must be more than a nominal quantity (reasonable), yet it should not exceed the amount the buyer is fairly certain to order.

   c. Suitability:

This type of contract is primarily suited for work known to be needed during a specified contract period but for which the exact time and quantity is indefinite. A good example is periodic repairs on buildings or equipment where history reflects a virtual certainty of need but flexibility is required.

2. Requirements
   a. Essential elements:

   Use where it is impossible to determine in advance the precise quantities of supplies or services that will be needed during a definite contract period.

   Method of ordering work must be stated as well as minimum/maximum orders allowable during a specified time period.

   The buyer is obligated to order from the successful supplier, and no other source, all supplies or services described in the contract during the stated contract period.

   Contract contains estimated quantities used for offer evaluation and clearly states that the buyer is not obligated to place any minimum orders (obligation is solely based on need required to be filled by that supplier).

   b. Limitations:

   Funds are obligated by each order and not by contract.

   Provides flexibility in quantity and delivery schedule as orders are placed only after need materializes.

c. Suitability:
Primarily suited for procuring services known to be a requirement but for which the exact timing and quantity are not predictable, such as emergency road service for vehicles.

COST-REIMBURSEMENT FAMILY OF CONTRACTS

Cost-reimbursement types of contracts provide for payment of allowable incurred costs to the extent prescribed in the contract. These contracts establish an estimate of total cost for the purpose of obligating funds, determining the fee, and establishing a ceiling that the supplier may not exceed (except at its own risk) without a modification to the contract. Cost-reimbursement types of contracts are suitable for use only when uncertainties involved in contract performance do not permit costs to be estimated with sufficient accuracy to use any type of fixed-price contract. Cost-reimbursement contracts are commonly used when the contract is for performing research or preliminary exploration or study and the level of effort required is unknown, when the contract is for development and testing, and when contracting with educational institutions. A cost-reimbursement contract should be used only when the supplier's accounting system is adequate for determining costs applicable to the contract, appropriate surveillance during performance will provide reasonable assurance that efficient methods and effective cost controls are used, and a cost-reimbursement contract type is likely to be less costly than any other type. In addition, buyers should use cost-reimbursement contracts when it is impractical to obtain supplies or services of the kind or quality required without the use of a fixed-price type of contract.

Cost–No Fee or Cost Sharing

A cost contract with no fee and a cost sharing contract are cost-reimbursement contract in which the supplier receives no fee. A cost contract with no fee may be appropriate for research and development services, particularly with nonprofit educational institutions or other nonprofit organizations, and for facilities contracts. If the buyer and supplier split the cost, cost sharing should be apportioned on some reasonable basis, such as a straight percentage split of all allocable costs. More commonly, cost sharing is based on reimbursement by the buyer of incurred direct labor, material, and equipment costs, with the supplier covering all indirect costs, including overhead and G&A.

Types of Contracts and the Analytical Methods Appropriate to Each 27

Cost-Plus-Fixed-Fee

A cost-plus-fixed-fee contract is a cost-reimbursement contract that provides for an estimated target cost and payment to the supplier of a negotiated fee that is fixed at the inception of the contract. The fixed fee does not vary with actual cost, but may be adjusted as a result of changes in the work. This type of cost-reimbursement fee-bearing contract provides the least amount of cost incentive for the supplier and should be avoided in favor of other fee-bearing contract types, if possible.

Cost-Plus-Incentive-Fee

A cost-plus-incentive-fee contract provides for a target cost, a target fee, a minimum and maximum fee, and a fee-adjustment or share ratio. Under this type of contract, the buyer reimburses the supplier for all actual allowable costs. Then, the buyer applies the fee-adjustment ratio. If the actual allowable costs exceed the target costs, the supplier's fee is less than the target fee and vice versa. In other words, the lower the supplier's costs, the higher the supplier's fee. Regardless of cost incurrence, the contract will not earn more than the maximum fee or less than the minimum fee. Performance incentives may be applied to the contract arrangement to supplement the cost incentive, if appropriate. The cost-plus-incentive-fee contract is suitable for use in development and test programs when cost (and performance) incentives are likely to motivate the supplier. The Wayne Corporation II case study expands The Wayne Corporation case, which was used to illustrate the fixed-price-incentive-fee contract.

The Wayne Corporation II

Assume The Wayne Corporation bitterly opposed using a fixed-price-incentive contract because of the risk inherent in the contract. The Wayne Corporation proposes in lieu thereof to perform under a cost-plus-incentive-fee contract with the following characteristics:

Target cost	$209,000
Target fee	11,000
Maximum fee	16,000
Minimum fee	5,500
Share ratio	$90/_{10}$

Given this information, let us determine the range of incentive effectiveness (RIE).

Discussion

The range of incentive effectiveness is the cost range within which the share ratio operates and the buyer and supplier will split the cost of the underrun or overrun. At the lower end of the range, the share ratio becomes 100% (the buyer reaps the benefit of any additional underrun) and the supplier receives the maximum fee agreed to. At the upper end of the range, the share ratio again becomes 100% (the buyer incurs the cost of any additional overrun) and the supplier receives the minimum fee agreed to.

The lower end of the RIE is computed by using the following formula:

$$\text{Target Cost} - \frac{(\text{Maximum Fee} - \text{Target Fee})}{(\text{Supplier Share})}$$

$$\text{Or: } \$209,000 - \frac{(\$16,000 - \$11,000)}{(.10)}$$

$$\text{Or: } \$209,000 - \$50,000 = \$159,000$$

The upper end of the RIE is computed by using the following formula:

$$\text{Target Cost} + \frac{(\text{Target Fee} - \text{Minimum Fee})}{(\text{Supplier Share})}$$

$$\text{Or: } \$209,000 + \frac{(\$11,000 - \$5,500)}{(.10)}$$

$$\text{Or: } \$209,000 + \$55,000 = \$264,000$$

To check the result, the reader can say: How much cost must I subtract from the target cost in order to increase the fee from $11,000 (target fee) to $16,000 (maximum fee)? The answer is $50,000, since $50,000 × .10 = $5,000. Or, the reader can say: How much cost must I add to the target cost in order to decrease the fee from $11,000 (target fee) to $5,500 (minimum fee)? The answer is $55,000, since $55,000 × .10 = $5,500.

A graphic representation of this cost-plus-incentive-fee contract is shown in Figure 2-4.

Cost-Plus-Award-Fee

The cost-plus-award-fee contract is a cost-reimbursement contract that provides for a fee in addition to a base fee fixed at the contract's inception. The award fee is intended to motivate the supplier toward excellence in such areas as quality, timeliness, technical ingenuity, and cost-effective management. The amount of the award fee to be paid is determined by the supplier's performance in terms of the criteria stated in the contract. Criteria are

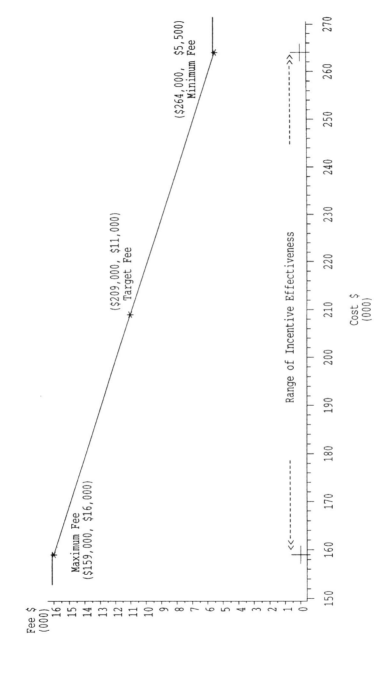

FIGURE 2-4.

generally of a subjective nature. The cost-plus-award-fee contract is suitable for use (1) when the work to be performed is such that it is neither feasible nor effective to devise predetermined objective incentive targets applicable to cost, technical performance, or schedule; (2) when the likelihood of meeting a purchasing objective will be enhanced by using a contract that effectively motivates the supplier toward exceptional performance and provides the buyer with the flexibility to evaluate both actual performance and the conditions under which it was achieved; and (3) when any additional administrative effort and cost required to monitor and evaluate performance are justified by the expected benefits.

"HYBRID" CONTRACTS

Time-and-Materials

This type of contract is commonly used in contracting for repair and maintenance work as well as professional services (including consulting). It provides for acquiring supplies or services on the basis of direct labor hours and materials. The direct labor hours are charged at specified fixed hourly rates that include wages, overhead, general and administrative expenses, and profit. Materials are charged at cost, including, if appropriate, material handling costs. A time-and-materials contract should be used when it is not possible to estimate accurately the extent or duration of the required work or to anticipate costs with any reasonable degree of confidence. Because time-and-materials contracts provide no particular cost incentive to the supplier, they should be used only when no other contract type is suitable and only if the contract establishes a separate ceiling amount for labor, beyond which the fixed hourly rates will be reduced by deleting the profit included therein (in order to avoid a cost-plus-a-percentage-of-cost situation).

A typical schedule for a time-and-materials contract is shown in Table 2-4.

Labor-Hour

Labor-hour contracts are a variation of the time-and-materials contract, differing only in that materials are not supplied by the supplier.

Combination Contracts

The buyer will be confronted with many situations that call for flexibility in structuring contracts. It is not uncommon to be confronted with a situation that requires considering a combination firm-fixed-price/requirements/time-and-materials contract, particularly where a contract must, of necessity, incorporate both service and construction work. Combination/composite contract

TABLE 2-4   Schedule of Time-and-Materials Services

Schedule					
0003	Provide Professional Education and Training Services in accordance with the Statement of Work. The clause in this contract entitled "Time and Materials" explains this type of contract and its operation.				
Item No.	Supplies/Services	Estimated Quantity	Unit	Unit Price	Amount
0003A	Professional educator hours	500	HRS	$ 150	$75,000
0003B	Materials at cost (estimated)	XX	XX	XXXX	$10,000
	Total price for contract (line item 0003)	XX	XX	XXXX	$85,000

thinking has logically spread into the supply and service contract business in general. Many contracts call for work "as needed," which means the contract has a variety of delivery and cost considerations. Combining contract types in a single contract document for associated services is practical because it reduces the number of formal solicitation packages that have to be prepared, the solicitation effort, and the resulting contract documents. For instance, one contract can contain a combination of a firm-fixed-price and time-and-materials work.

Let us assume for illustrative purposes that the buyer needs building maintenance and repair services. Assume further that the building maintenance service requirements are firm and definite and that substantial historical experience is available to establish the scope and level of required maintenance. The building repair services requirements are sporadic and fluctuate to a large degree. Moreover, the repair services are such that they must be performed by construction tradesmen. The buyer knows the exact amount of maintenance work based on historical data, an approximate amount of repair work based on historical data, that the nature of the maintenance and repair work can be defined, and that sound estimates for maintenance work and less-firm estimates for repair work can be established. The buyer does not know when the repair work will be needed, what specific types of repair will be ordered, or the exact number of buildings to be repaired. The buyer should select a combination firm-fixed-price and time-and-materials contract. Selecting this combination is appropriate because the maintenance work can be adequately defined in a firm-fixed-price line item; however the quantity and frequency of repair work cannot be defined. In addition, the fact that repair work must be done by construction tradesmen provides an added incentive to separate it from the maintenance work. It is not possible to develop firm-fixed-unit priced discrete work line items for this work.

AGREEMENTS AND PRECONTRACTUAL INSTRUMENTS

Blanket Agreement

A blanket agreement is a written instrument of understanding that contains terms and conditions applying to future orders (calls) between the parties during its term; a description, as specific as practicable, of supplies or services to be provided; and methods of pricing, issuing, and delivering future orders (calls) under the blanket order.

A blanket agreement is not a contract. Blanket agreements may be used if there is a wide variety of items in a broad class of services or goods that are generally purchased, but the exact items, quantities, and delivery requirements are not known in advance and may vary considerably. Blanket agreement may also be used to avoid writing numerous purchase orders. A ceiling must be established for the total instrument as well as for each order (call).

Letter Contract

A letter contract is a written preliminary contractual instrument that authorizes the supplier to begin immediately performing services. Their use should be strictly discouraged.

SUMMARY

Negotiating price is meaningless unless that price is related to the type of contract to be awarded. A price that appears fair and reasonable on a firm-fixed-price basis may be unreasonable on a cost-reimbursement contract. Like price, contract type is generally negotiable. This fact requires the buyer to prepare multiple pricing and contract type objectives for negotiation.

Risk assessment is critical in selecting the type of contract. The objective is to select a type of contract with the greatest incentive for efficient and economical performance. Different types of contracts represent varying degrees of risk to the buyer and supplier.

There are other factors to consider when selecting an appropriate contract type. These factors, which vary in their degree of importance from purchase to purchase are supplier motivation, the degree of competition in the marketplace, the degree of sophistication in the supplier's accounting system, the method of procurement to be used in selecting the supplier and negotiating the contract, and the administrative work load necessary to administer the contract.

3

Price Comparison Methods and How to Use Them

DIFFERENT CONDITIONS REQUIRE DIFFERENT ANALYTICAL METHODS

The conclusion that a price is fair and reasonable must be based on some form of price or cost analysis. How detailed this analysis will be depends on the dollar value and the nature of the product or service being purchased.

PRICE ANALYSIS

Lee, Dobler, and Burt (1990, Pg 253) assert that "Some form of price analysis is required for every purchase." Although this is true, price analysis takes on special importance when it alone is employed (without cost analysis). Price analysis alone is generally used for low-dollar purchases; for most competitive purchases, even though of a large dollar value; for purchases based on existing catalog or market prices; and for purchases of items or services for which regulated prices exist (regulated utility services).

To perform price analysis, the buyer must have a base or reference to which the quoted price can be compared. That basis for comparison must itself be reasonable. Then the buyer must ensure that the quotation and the base are truly comparable; that the comparison is "apples to apples." That is why the comparison of competitive quotations is such an effective method of price analysis. If the buyer can be reasonably assured that the items are comparable and, presuming that the firms involved are really competing with one another, then the lowest price submitted will be reasonable.

In performing any price analysis, then, comparability is the key. One must consider the following:

34 Cost/Price Analysis: Tools to Improve Profit Margins

- quality of the items for which prices are being compared;
- quantities involved in the sale;
- delivery conditions (F.O.B. origin versus F.O.B. destination); and
- market conditions (for some items, tomorrow's price may be different from yesterday's, often by a wide margin).

Given a reasonable base for comparison, and allowing for the comparability issues listed above, price analysis techniques will determine whether a price is fair and reasonable.

Methods of Price Analysis

Many methods of price analysis are available. Selecting the right method depends on the specific features of the purchase. In many instances, a combination of methods is best. The following methods are among the most commonly used.

Comparison With Competing Offers on the Instant Purchase

This is generally considered a primary method of price analysis. It is nothing more than the act of seeing which price quote is the lowest among those received. Unless there is some doubt about the adequacy of competition, this method is generally considered a conclusive judge of price reasonableness. The buyer must be sure, however, that the prices compared are submitted on the same basis and that factors such as transportation charges to be paid by the buyer (when delivery is to be f.o.b. origin) have been considered.

Comparison With Established Catalog Prices

In the absence of price competition this method of price analysis should be considered a primary method of comparison. Because many suppliers publish prices for items that are regularly offered for sale, quotations can be compared to those published prices—but caution is required. First, the buyer must make sure that the catalogs represent actual prices that are now or were recently being charged. (Catalog prices are frequently discounted for government agencies and large corporations, which are generally considered "most favored customers" because of their large volume of business). Second, the buyer must be sure that the price listed applies to an item sufficiently similar to the required item to provide a sound basis for comparison. Third, the buyer must be sure that the catalog or price list applies to the same class of trade. If one item at a time is being purchased, comparison to the Sears or Montgomery Ward catalog price is appropriate. But if the purchase is for wholesale quantities, the Sears' or Ward's prices are not appropriate since they do not reflect normal trade or quantity discounts.

The "Bum Company" case study illustrates price analysis by comparison against established catalog prices.

Bum Company

The Bum Company has been manufacturing a series of similar products for several years. Two of the items required by the Steer Company were priced in the Bum Company catalog as follows:

Product	Catalog Price
Electric Motors	$196.75
Adjustable pumps	$174.25

The Bum Company offered to supply the Steer Company with 2,800 motors at a unit price of $195.75 and 750 pumps at a unit price of $174.25. Ms. Waters, the buyer for the Steer Company, was given the opportunity to review the quoted prices to determine if they were fair and reasonable in relation to the catalog prices.

Ms. Waters felt fairly sure the Pump price was appropriate. She knew the Steer Company was not a major customer of the Bum Company and that Bum did not offer anyone a discount from list. She was unsure about the Motor price, particularly since the offer was to sell below catalog price. She wondered what the basis was for the discount from list and whether other customers ordering similar amounts had received equal or greater discounts. What should Ms. Waters do under the circumstances?

Discussion
Ms. Waters could ask the Bum Company how they arrived at the discount for Motors and ask Bum to explain their policy for discounting from list. Ms. Waters could also suggest that the quantity being ordered by Bum might warrant a quantity discount greater than the one offered. If Ms. Waters was not satisfied with the Bum explanation, she might consider pricing the Motors against similar pumps in the market. Lastly, if Ms. Waters felt Bum was "gouging" the Steer Company, she might suggest an alternative source be attempted.

Comparison With Established Market Prices
Like the comparison with established catalog prices, this method of price analysis should be considered a primary method in the absence of price

competition. Many items are regularly traded at prices that fluctuate over short intervals. Catalogs of prices for these items are not published because changes occur too rapidly. But if the buyer can establish the price range in which sales are being made to the general public (through trade journals or other sources), that range can be used as a basis for comparison.

Comparison With Prices Set by Law or Regulation

This primary method of price analysis is appropriate when dealing with regulated utility services. Regulated utilities (electric companies and the like) are required to seek the approval of their regulatory commissions before adjusting utility rates. These public regulatory commissions are zealous protectors of consumers' interests and grant increases in rates rather sparingly. The regulatory commission's approved rates will be published and available to the buyer. The buyer, however, must be assured of the fact that the classification being used by the utility is correct. Many electricity tariffs (rate schedules) will have different rates for heavy and light users; for commercial, industrial, and governmental users; and for nonprofit users. The buyer must assure the lowest rate available is applied. In addition, the buyer must recognize that his or her organization may have an opportunity to impact the rates by "intervening" at the public utility hearings. Finally, organizations can sometimes establish their own rate categories. These opportunities should be explored if appropriate.

Comparison With Current Prices Paid for the Same or Similar Items, Past Prices Paid for the Same or Similar Requirements, and Past Offers

This method of price analysis should be considered only after one or more of the primary methods has been attempted. If the price quoted is the same or less than that recently paid for the same or similar items, the current quotation is likely to be reasonable. Buyers should be careful, however, to ascertain that the historical prices to which a comparison is made were themselves adequately analyzed for reasonableness. Further, buyers should take into account price trends (up or down) caused by market or economic conditions, rapidly fluctuating commodity prices, or other factors. Often the buyer will have a basis for comparison in the form of prices for a similar item. If the buyer can, through some method of price analysis, determine what the difference in price should be between the item being purchased and the one for which prices are available, that difference can be used as an adjustment in order to arrive at a valid comparison. Price history information can be obtained from the records of the Purchasing Department, the requisitioning activity, or if necessary, from other buying offices. Again, buyers must be sure that the price history applies to the same items under the same conditions.

Price Comparison Methods and How to Use Them 37

The "Audio Company" case study illustrates the danger of basing price reasonableness solely on a comparison with previous prices paid.

Audio Company

The Nadir Company has awarded a series of contracts for portable radios (used by its security forces) to the Audio Company for the past four years. There is a need to purchase an additional 200 units. These radios have excellent distance capabilities, the moisture-proofing is far superior to any other make on the market, and the maintenance required on the sets is practically nonexistent. Ms. Melody Samuals, the buyer, has compiled a purchasing history on the Audio Company (Table 3-1).

Ms. Samuals notes that each contract has resulted in a declining price. This certainly makes the radio more attractive, and further, allows her company to stretch its budget. She must determine if this is a fair and reasonable price.

Discussion

Ms. Samuals may or may not have a fair and reasonable price. A comparison with previous prices paid may provide a valid basis for assuming a fair price if the previous prices were determined fair based on some form of price cost analysis. Ms. Samuals would need to look at those previous purchase transactions to ensure that such was, indeed, the case. Another troublesome matter is that prices have gone down over time, even for purchases where quantities purchased were comparable. Ms. Samuals may want to go beyond this method of comparison and look at the Producer Price Index for this type of item. Although prices generally go up (with inflation), prices for some items go down. The price index trend for this or similar items should be investigated. A potential explanation for this situation may be that the Audio Company is experiencing a steep learning curve and is saving labor hours in production. At any rate, Ms. Samuals needs to discuss the matter with the Audio Company to seek an explanation for the price behavior.

TABLE 3-1

Purchase	Quantity	Unit Price
1st	100	$175
2nd	150	$170
3rd	200	$160
4th	225	$155
Proposed	200	$150

Comparison With Producer Price and Other Market Indexes

This technique, although it is generally considered a secondary method of price analysis, is very powerful and should be considered even when other more obvious comparative methods are available. This method, in conjunction with comparing similar purchases, is often used in order to make prices from different points in time comparable with respect to a specific point in time. There are numerous indexes available for using in the following analytical procedures:

- Adjusting previous prices for the effects of inflation from the purchase date to the present.
- Adjusting current prices for the effects of inflation from the present to the previous purchase date.
- Extrapolating current prices into the future based on the assumption that future price increases will mirror the past. This is generally done through forecasting index numbers.
- Comparing prices paid in two successive contracts with increases in market prices during the same period.
- Comparing price increases in one commodity against price increases in another commodity during the same time period.
- Comparing general growth in prices (Consumer Price Index) with growth of specific items or commodities (Producer Price Index).

Although the Consumer Price Index and the Producer Price Index (both monthly publications available from the U.S. Department of Labor, Bureau of Labor Statistics (BLS)) are the two most commonly used indexes in price and cost analysis, there are several other publications which may be of interest to buyers. The BLS also publishes wage and benefit changes resulting from collective bargaining settlements and unilateral management decisions, statistical summaries, and special reports on wage trends in its monthly publication *Current Wage Developments*. The BLS also publishes articles on labor force, wages, prices, productivity, economic growth, and occupational injuries and illness in its monthly publication entitled *Monthly Labor Review*. Still other helpful BLS publications include the *Area Wage Surveys* published throughout the year. These surveys include office clerical, professional, technical, maintenance, custodial, and material movement occupations. The U.S. Departments of Commerce, Interior, and Energy also publish specialized information that may be helpful in analyzing labor and material price trends. Many commercial and industrial publications have helpful pricing information that can also be used to analyze labor and material trends. Many of these are industry specific.

The "Index Company" case study illustrates how an index from the

Producer Price Index can be used to extrapolate past prices to the present. In a slight adaptation of this technique, the "Moving Average Company" case study illustrates how to project material prices using a moving average of historical prices.

Index Company

Uri Tinker II is a buyer for the Thinker Company. He was assigned the task of determining whether $120 is a fair and reasonable price to pay for a "Thinkmajig." He determines Thinkmajigs are categorized as "Refined Metal Products" in the Producer Price Indexes. He extracts the price index history shown in Table 3-2 from that publication (1987 base).
In 1988 he paid $100.00 for a Thinkmajig. This is now December 31, 1993. What is a reasonable price to pay?

Discussion
This case requires the buyer to escalate an old price to the present using the price index increase during the intervening period as the basis for inflation. The percent of increase is determined by dividing the 1993 index by the 1988 index. The result is 119.5/103.1 = 1.16. A reasonable price to pay is $100.00 × 1.16 = $116.00. If the Thinkmajig had been purchased in 1987 for $100, the expected price would have been $119.50.

Moving Average Company

Yorel Warg was a buyer for the Pedal-Metal Construction Company. She was assigned the task of forecasting the price of sheet metal for the first quarter of 1993. She decided to accomplish this by computing four-quarter moving averages, projecting a four-quarter moving average for the first quarter of 1992, then extrapolating an unadjusted price per sheet. See the information shown in Table 3-3.
What would be a reasonable projection for the price during the 1st quarter of 1993?

TABLE 3-2

1987	1988	1989	1990	1991	1992	1993
100.0	103.1	106.2	108.2	109.8	114.3	119.5

TABLE 3-3

Year Average	Quarter	Price	Four-Quarter Moving
1991	1	$20	-
1991	2	$28	-
1991	3	$29	-
1991	4	$15	$23.00
1992	1	$22	$23.50
1992	2	$31	$24.25
1992	3	$33	$25.25
1992	4	$17	$25.75

Discussion

To solve this problem, Yorel must work backwards by projecting the four-quarter moving average into 1993 and then solving for the price mathematically. If she plots the four-quarter moving average on graph paper, she will observe the average moves up about .875 (on the average) each year. If she adds .875 to the moving average for the fourth quarter of 1992, she can derive a value of $26.62. Armed with that data, she can algebraically solve for the 1st Quarter 1993 price as follows:

$$\text{Moving Average for 1st Quarter 1993} = \frac{\text{(Prices for 2nd Quarter 1992, 3rd Quarter 1992, 4th Quarter 1992, and 1st Quarter 1993)}}{4}$$

$$\$26.62 = \frac{(\$31 + \$33 + \$17 + X)}{4}$$

$$106.48 = \$81 + X$$

$$X = \$25.48$$

Comparison With Cost Estimating Relationships, to Include Rough Yardsticks and Parametric Relationships

This comparison technique is considered by many to be a powerful secondary method of price analysis. Cost estimating relationships are defined as relationships between cost and an item or service characteristic. Typical cost estimating relationships include the following:

- The cost of construction based on floor space, roof surface area, and wall surface.
- The cost of gears based on gear net weight, percentage of scrap produced

by the gear, inches of teeth cut into the gear, the hardness of the gear, and/or the envelope of the gear.
- The cost of trucks based on truck empty weight, truck gross weight, horsepower, number of driving axles, and loaded cruising speed.
- The cost of passenger cars based on curb weight, width of wheel base, square feet of passenger space, and horsepower.
- The cost of turbine engines based on dry weight, maximum thrust, cruise thrust, specific fuel consumption, bypass ratio and inlet temperature.
- The cost of reciprocating engines based on dry weight, piston displacement, compression ratio, and/or horsepower.
- The cost of sheet metal based on net weight, percentage of scrap, number of holes drilled, number of rivets placed, inches of welding, and/or volume of envelope.
- The cost of aircraft based on empty weight, speed, useful load, wing area, power, and/or landing speed.
- The cost of diesel locomotives based on horsepower, weight, cruising speed, and maximum load on standard grade at standard speed.

The "and/or" in these explanations suggest that cost estimating relationships can be based on more than one independent variable (the "based on" factors). Although the simplest cost estimating relationships are linear (indicating a straight line on a graph of the independent and dependent variables), some relationships are curvilinear (other than a straight line). Such relationships are best dealt with by using computers. Actually modern computer programs are helpful in developing all types of cost estimating relationships and are even more important when forecasting dependent variables (the "costs" "based on" the independent variables).

When developing a cost estimating relationship, it is necessary first to designate and define the dependent variable (the factor that is influenced or caused by the independent variable). The dependent variable is generally cost or man-hours.

After designating the dependent variable, one must select item characteristics to be tested for estimating the independent variable. The independent variable is the factor that influences or drives the dependent variable. Finding that independent variable is not always easy. The buyer will need to draw on his or her personal experience, the experience of others, and published information. In selecting the independent variable, the buyer should consider only factors that are (1) readily available in a statistically usable form; (2) quantitatively measurable; (3) related to performance characteristics of the item or system being explained by the cost estimating relationship.

After selecting the dependent and independent variables, the buyer will need to collect data concerning the relationship between the two. Data for at least five points (often years) should be collected. The data should be checked to assure that it is relevant, comparable, and free of unusual elements.

After collecting the data, the buyer will need to examine the relationship between the independent and dependent variables. This examination can be something as simple as graphing the data or as complex as running the data through a computer regression analysis program. The purpose is to test the degree of relationship between the variables. A high correlation coefficient between the variables usually indicates that the independent variable will be a good predictive tool.

After examining the relationship, the buyer will need to determine the relationship that best predicts the dependent variable. This generally requires testing more than one independent variable against the dependent variable.

Once the best relationship is identified, the model or formula resulting from the effort is saved for use as a forecasting tool.

The "Estimating Relationship Company" case study illustrates how a cost estimating relationship can be established between office building construction cost and square footage of work space.

Estimating Relationship Company

Sam Dillon is a construction buyer for Mercy Hospital. He is about to issue a solicitation for a new Proton Cancer Treatment Facility. He has an in-house estimate from his Engineering Department, but he wants to test that estimate before he goes out on the street. He thinks he can develop a good estimating relationship by gathering historical cost of construction data and relating that to square footage of construction for several other cancer treatment facilities within the state. He is hesitant to go outside the state since their costs may not be in line with his state. Sam calls his other hospital purchasing contacts and collects the numbers shown in Table 3-4.

Sam needs to know whether $175,000,000 (the in-house estimate) is realistic for a treatment facility of 29,000 square feet.

TABLE 3-4

Hospital Reporting	Treatment Facility Cost	Square Feet
Seaside County	$166,500,000	28,000
City of Bellevue	$165,000,000	27,000
State Hospital	$168,000,000	28,600
Land of Parks	$160,500,000	24,440
Skinner Hospital	$163,000,000	26,000

Discussion

Since Sam has identified his independent and dependent variables, he can plot the variables. Such a plot would approximate the one shown in Figure 3-1.

By running a "line of best fit" through the five given data points, Sam can readily determine that the Engineering Department's estimated facility cost does not fall on the extension of that line. Sam would, in this circumstance, want to ask the Engineering Department to explain why their estimated cost was so much higher than recently paid by other hospitals for similar type facilities.

Comparison With In-house Estimates

If the requisitioner has developed a competent cost estimate for the item or service to be acquired, the quoted price can be compared with that estimate. Care should be exercised to ensure that the estimate is a sound base for comparison and accounts for all factors that will affect the price. The estimate should have been prepared when the PR was submitted, without knowledge of the quotes received. The estimate is not to be based on the amount of program funds available but on the current market prices for the goods and services being requisitioned. Detailed, bottom-up estimates are particularly important in purchasing services, particularly construction services. A detailed in-house estimate will be very helpful to the construction buyer for comparison with the low bidder's working papers whenever there is some question about whether the low bidder has overlooked some work. By doing a side-by-side comparison of the low bidder's working papers and the in-house estimate, serious estimating errors can be discovered. The prudent course of action for the buyer in such an eventuality is to permit the low bidder to withdraw his bid and proceed on to the next bidder (who would generally be subjected to a similar type of comparison). If, of course, the low bidder had bid in line with the in-house estimate, there would be no need for a side-by-side comparison.

In-house estimates are also essential in performing cost analysis, particularly when competition is limited. When a potential supplier submits a cost proposal in response to the solicitation, the buyer generally relies on his or her own engineer's estimate of required man-hours, material quantities, and equipment quantities to analyze the supplier's cost proposal. In many cases, the labor rates, material prices, and equipment rates included in the in-house estimate are good guides to follow. Generally, however, these rates (and the accompanying indirect rates) are subject to additional verification by reviewing the supplier's books and records or performing a market rate analysis.

The "Dynamic Trio Companies" case study illustrates a procurement situation that required price analysis by comparing competing offers against

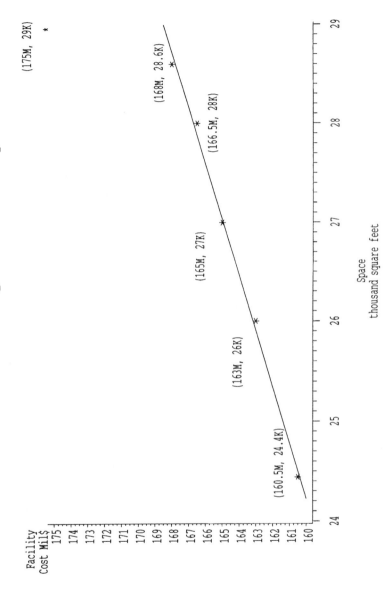

FIGURE 3-1.

each other and the in-house estimate. The case study also illustrates the shortcomings of price analysis in the face of a sole-source procurement.

Dynamic Trio Companies

Mr. Ira Bateman, buyer for Gorham City, has received a requisition for engineering services. The in-house estimate is $250,000. Bateman prepares a solicitation, which he sends to all small and small-disadvantaged firms registered with the City. Assume three firms respond with the following results:

Services R'Us $240,800
Small But Mighty $245,000
Robin The Hood $252,000

Bateman is not sure whether he has effective price competition. Moreover, he is not sure whether he should make an award to Services R'Us without further discussion. Finally, he isn't sure what price analytical techniques can assure a fair price.

Discussion
Bateman may have obtained effective price competition, particularly if his solicitation contained a good statement of work and all three offerors understood what they were bidding on. More commonly purchasing engineering services requires further discussions to clarify scope. This procedure is necessary to assure commonality of thought among all parties. Still, if he has confidence in his estimate, Bateman may well conclude discussions would only add to the length of the transaction and decide to award without substantive discussions. Beyond comparisons with competing offers and the in-house estimate, additional price analytical measures may be unnecessary.

Assume that instead of the above results, Bateman received the following offers:

Services R'Us $225,000
Small But Mighty $330,000
Robin The Hood $320,000

Under this circumstance, Bateman is even less sure whether he should rely on a comparison with competing offers and comparison against the in-house estimate. He wonders whether some degree of cost analysis may not be necessary.

Discussion

This situation would probably prompt Bateman to go beyond two price comparison techniques. He would seemingly be interested not only in why Small But Mighty and Robin The Hood bid so high, but also why and how Services R'Us bid at 90 percent of the in-house estimate. Partial cost proposal information, needed for comparison with the in-house estimate, would be necessary for Bateman to determine if the City made a mistake on its estimate or whether the three offerors made mistakes in their estimates.

Assume that it is 90 days later. Services R'Us is under contract to perform the engineering services. The City engineer requiring the services has identified some work related to the work currently under contract, that needs to be completed simultaneously with the work under contract. The engineer is prepared to satisfy the City's requirements for justification of an out-of-scope change to the Services R'Us contract so that Bateman can modify the contract. What must Bateman do in this circumstance?

Discussion

This out-of-scope change to the Services R'Us contract requires a sole-source negotiation based on a cost proposal submittal by Services R'Us. Bateman must conduct not only price analysis (not very helpful in this case) but also cost analysis of the Services R'Us cost data in order to arrive at a negotiation position. Bateman must also conduct some form of profit analysis in order to arrive at a reasonable prenegotiation profit position. These techniques are covered in the discussion of cost and profit analysis.

Comparison With Values Determined by Value and Visual Analysis

These techniques are generally considered tertiary or "auxiliary" price analysis techniques. Value analysis is the task of determining why seemingly similar products should be priced differently. The technique helps the buyer understand the reason for price differences between past buys and present offers. Buyer value analysis generally concentrates on the utility and aesthetic qualities of similar items in order to derive opinions of respective value. The analysis normally takes place in two stages. In the first stage the buyer lists the functions required and compares the required functions to those of alternative products. In doing this, the buyer assumes that an item with a lower use value should have a lower price (an assumption that may not be supported in fact). In the second stage, the buyer identifies and compares the aesthetic functions to those of alternative products. Upon completing the second stage, the buyer often finds that the price differentials, if any, are supported by the aesthetic differences rather than the use differences. Commercial and industrial entities are generally more interested in use than aesthetic value.

Visual analysis relies on visual inspection of the item (or alternatively, the representations of the item in drawings) to develop a rough estimate of the value. Because it concentrates on obvious, external features of an item, it should be used only for "ballpark" analysis. However, visual analysis can prevent mistakes and oversights and lead to questions about offered prices. The buyer who bought the now infamous "$600 Hammer" would probably have benefitted from this analytical technique.

PLACING OFFERS ON THE SAME COMPARATIVE BASIS

In conducting price analysis to arrive at source-selection decisions, it is critical for the buyer to place all offers on the same basis. Moreover, it is important that the cost of the purchasing process itself be considered when choosing between a multiple and a single award. When quotations are obtained on related items such as various MRO items, small hardware items, equipment parts, or office supplies, the buyer should stipulate in the solicitation that the buyer reserves the right to award on an all-or-none basis; that is, the buyer may purchase from the offeror who submits the lowest total price for all items, rather than issue a purchase order to each supplier on the basis of the lowest quotation on each item. Purchasing on the basis of lowest total cost may afford savings since the cost of writing and administering multiple purchase orders and making multiple payments may be rather expensive. Many buyers estimate the cost of issuing an order and paying the invoice for a small order at $250.

This amount is generally used as an evaluation factor when considering awarding multiple orders from a single solicitation. Buyers in a given organization should evaluate their own organization's buying costs before performing this type of analysis. Many organizations have administrative costs that far exceed $250 per award.

In the example in Table 3-5, award in the aggregate should be made to Supplier B because that will result in a total savings, even though B was not low on each item.

To determine whether purchases should be made based on lowest total cost, it is necessary to ascertain the administrative cost of the method of purchase.

It may not always be advantageous to award "all or none." Suppliers should be advised that the buyer reserves the right to award (or not award) on that basis after evaluating vendors' quotations.

A supplier also may add a condition that award will be accepted only on

48 Cost/Price Analysis: Tools to Improve Profit Margins

TABLE 3-5

Item No.	Suppliers			Low Quotation
	A	B	C	
1	$125	$130	$133	$ 125
2	$150	$144	$147	$ 144
3	$148	$143	$140	$ 140
Totals	$423	$417	$420	$ 409
Cost to Issue Purchase Order(s)		$250		$ 750
Total Cost		$667		$1159

an all-or-none basis. The reason for doing so is simple. Costs for delivering the three items together may be much less than the total cost for delivering the three individually. An all-or-none award means that the supplier too has only one order to process.

The example in Table 3-6 shows a case lacking an all-or-none qualification by a supplier, in which award should be made on an item-by-item basis to the lowest quoter on each item. Although there are three orders to process, the total cost to the buyer will be less than if the order had been placed with the firm quoting the lowest overall price for all three items.

The "Ms. Betty Pool" case study further illustrates the method of arriving at a bottom-line price comparison, considering the cost of making multiple awards.

Ms. Betty Pool

Ms. Betty Pool, a buyer working for the Electric Utility Company, has received a requirement to purchase 750 small motors and 250 small genera-

TABLE 3-6

Item No.	Suppliers			Low Quotation
	A	B	C	
1	$ 700	$ 900	$ 985	$ 700 (A)
2	$ 710	$ 420	$ 795	$ 420 (B)
3	$ 798	$ 890	$ 460	$ 460 (C)
Totals	$2,208	$2,210	$2,240	$1,580
Cost to Issue Purchase Order(s)	$ 250			$ 750
Total Cost	$2,458			$2,330

tors for use by field maintenance forces. These were standard commercial items. After reviewing the purchase request, Ms. Pool issued the solicitation on an "any-or-all" basis. The solicitation contained a $250 cost per contract for making multiple awards. In response to the solicitation, five companies competed for motors and five companies competed for generators. The abstract of bids reflected the results shown in Table 3-7.

Ms. Pool isn't sure to whom she should make the award in light of these circumstances.

Discussion
Ms. Pool must consider the bottom line cost of making an award to the lowest bidder on both items combined, of making an award to the lowest bidder for each of the items, and of making an award to the lowest of all possible bidders. The latter computation would mean an award to Lazy for 150 generators, with the award for 100 generators going to Sneezy. The statistics for this would appear as in Table 3-8.

This investigation proves that three awards on an any-or-all basis presents the lowest cost alternative for the Electric Utility Company.

DISCOUNTS

In the "Ms. Betty Pool" case study, Sleepy and Bashful companies both offered prompt payment discounts. Although Ms. Pool did not evaluate the discount or base the award on that evaluation, she may well consider discounts in the evaluation when appropriate. Every buyer should assess any offered discounts and determine what impact they will have on the decision about who gets the order. Discounts fall within two categories: trade (favored customer) and term (prompt payment).

- *Trade (favored customer) discount.* This is a discount off the unit or total price, given by the supplier to its best customers. It will generally be indicated on the face of the quotation under the pricing information.
- *Term (prompt payment) discount.* This is a discount that the supplier gives for prompt payment. The solicitation should have a provision for the supplier to offer a discount at its discretion. This was the type of discount offered to Ms. Pool.

Astute buyers solicit or attempt to negotiate prompt payment discounts as a means of lowering the bottom line price. A prompt payment discount is a reduction in price on the condition that the buyer pays the bill within a certain number of days after receipt of an invoice. Prompt payment discounts can be a significant source of savings.

It is often to the supplier's advantage to take a smaller payment quickly

TABLE 3-7

Firm	Item	Qty.	Unit $	Total $	Item	Qty.	Unit $	Total $	Rmks.
Doc	Motors	750	$62.00	$46,500.00	Generators	250	$142.00	$35,500.00	
Sneezy	Motors	750	$57.00	$42,750.00	Generators	250	$136.25	$34,062.50	
Dopey	Motors	750	$55.75	$41,812.50	Generators	250	$137.50	$34,375.00	All or None
Sleepy	Motors	750	$56.00	$42,000.00	Generators	250	$138.00	$34,500.00	2/10 Net 30
Bashful	Motors	750	$55.80	$41,850.00	Generators	0	No Bid	No Bid	2/20, Net 30
Lazy	Motors	0	No Bid	No Bid	Generators	150	$134.55	$33,637.50	150 (Only) Generators

TABLE 3-8

No Split Award		
Dopey Motors	750 ea. × $55.75 =	$41,812.50
Dopey Generators	250 ea. × $137.50 =	$34,375.00
	Subtotal	$76,187.50
	One award	$ 250.00
	Total	$76,437.50
Split award (2 ways)		
Bashful motors	750 ea. × $55.80 =	$41,850.00
Sneezy generators	250 ea. × $136.25 =	$34,062.50
	Subtotal	$75,912.50
	Two awards	$ 500.00
	Total	$76,412.50
Split award (3 ways)		
Bashful motors	750 ea. × $55.80 =	$41,850.00
Sneezy generators	100 ea. × $136.25 =	$13,625.00
Lazy generators	150 ea. × $134.55 =	$20,182.50
	Subtotal	$75,657.50
	Three awards	$ 750.00
	Total	$76,407.50

rather than wait for a few extra dollars. Suppliers have to worry about their cash flow; they must have money on hand to pay their employees and to meet their obligations. Their accounts receivable, that is, money owed to them, cannot be used to pay those bills. If they do not have enough cash, they must often borrow and pay interest. Also, if they do not receive payments, they may have to go through the process of issuing a second invoice to ensure that the first one was received, which process of course costs money. For these reasons, even large companies may offer a discount to customers who pay quickly.

Prompt payment discounts are generally stated as a percentage off the stated price if payment is made within a certain number of days. The notice "2 percent, 20 days," means that if the invoice is paid within 20 days of the date it is received, the customer may deduct 2 percent from the total. On a $5,000 purchase, for example, a 2 percent discount means a savings of $100. Discounts may be offered on a sliding scale: "2 percent, 20 days; 1 percent, 30 days."

When written solicitations are used, the buyer should always check to see if the supplier has offered a prompt payment discount. When oral quotations are received, the buyer should inquire about such discounts. The buyer should, under certain circumstances, solicit prompt payment discounts and may, if organization policy permits, consider the discount in the price evaluation.

Because the time needed to process payments often exceeds 10 days for most organizations, discounts of less than 20 days should generally not be negotiated.

TRANSPORTATION AND DELIVERY CONSIDERATIONS

Another important impact on the bottom line for price comparison purposes is the delivery point (which determines the point at which ownership is transferred. Free on board (F.O.B.) point refers to the location at which the supplier delivers the supplies or materials to the buyer. The supplier owns the goods until they reach that point, and has responsibility for shipment and damage to that point. At that point, the buyer takes ownership of the supplies or material, and is responsible for damage and costs incurred after that point. The owner of the goods is responsible for determining and exercising control over the shipment of the goods. There are three major types of F.O.B. points in common domestic use.

"F.O.B. Origin" means free of expense to the buyer delivered on board the indicated type of conveyance of the carrier (or of the buyer, if specified) at a designated point in the city, county, and state from which the shipment will be made and from which line-haul transportation service (as distinguished from switching, local drayage, or other terminal service) will begin. When quotations have been requested on an F.O.B. Origin basis, the buyer will take possession of the item at the supplier's location, arrange for transportation, be responsible for loss or damage in transit, and pay any freight charges involved in getting it to the point of actual use. In such a case, the buyer must determine the lowest applicable freight charge between the vendor's location and the point of use. When specifying F.O.B. shipping point, the buyer should be sure to understand where that shipping point is. The sales office could be in Seattle, but the plant and shipping point in Miami. Shipping costs could be much more expensive than anticipated. Freight rates can be obtained from the traffic unit of the organization. This amount must then be added to the quoted price to arrive at a price for evaluation. Obviously, the same procedure must be used for all vendors' prices. Sometimes the supplier transfers title to the buyer upon delivery of the goods to the carrier, but the supplier prepays the shipping costs and charges the costs to the buyer. The buyer owns the goods during shipment, and is responsible for loss or damage en route.

"F.O.B. Origin, Freight Allowed" means free of expense to the buyer delivered on board the indicated type or conveyance of the carrier (or of the buyer, if specified) at a designated point in the city, county, and state from which the shipments will be made and from which line-haul transportation service (as distinguished from switching, local drayage, or other terminal service) will begin; and an allowance for freight, based on applicable published tariff rates between the points specified in the contract, is deducted from the contract price. In this instance, title is transferred to the buyer when the supplier delivers the goods to the carrier. The supplier reimburses the buyer for the transportation costs. Responsibility for loss or damage during transit is assigned to the buyer.

"F.O.B. Destination" means (i) free of expense to the buyer delivered on board the carrier's conveyance, at a specified delivery point where the consignee's facility (plant, warehouse, store, lot, or other location to which shipment can be made) is located; and (ii) supplies shall be delivered to the destination consignee's wharf (if destination is a port city and supplies are for export), warehouse unloading platform, or receiving dock, at the expense of the supplier. The buyer is not liable for any delivery, storage, demurrage, accessorial or other charges involved before the actual delivery (or "constructive placement" as defined in carrier tariffs) of the supplies to the destination, unless such charges are caused by an act or order of the buyer acting in its contractual capacity. If a rail carrier is used, supplies should be delivered to the specified unloading platform of the consignee. If a motor carrier (including "Piggyback") is used, supplies should be delivered to the truck's tailgate at the unloading platform of the consignee. If the supplier uses a rail carrier or freight forwarder for less than carload shipments, the supplier should ensure that the carrier will furnish tailgate delivery if transfer to truck is required to complete delivery to consignee. In F.O.B. Destination, ownership of the goods transfers to the buyer when they are delivered to the specified destination. The supplier pays the transportation costs and is responsible for loss or damage to the destination. This is the preferred method because the supplier must handle problems that occur before the goods are received by the buyer. When the F.O.B. point is at destination, an article that requests the supplier to "prepay and add" freight costs is inappropriate and should not be used.

Shipping terms for international transactions are generally based on common international usage as spelled out in the International Chamber of Commerce (ICC) booklet entitled *Incoterms*. These "INCOTERMS" are shown in Victor Pooler's[1] tabular presentation of those terms in Figure 3-2.

LIMITATIONS OF PRICE ANALYSIS

The "Janitorial Company" case study illustrates a situation where price analysis alone may not be sufficient to guarantee fairness and reasonableness

---

[1] Victor H. Pooler, *Global Purchasing: Reaching for the World*. New York; Van Nostrand Reinhold, 1992, p. 115.

54   Cost/Price Analysis: Tools to Improve Profit Margins

BUYER & SELLER RESPONSIBILITIES

B - BUYER RESPONSIBILITY
S - SELLER RESPONSIBILITY

	INCOTERM	BASIC RESPONSIBILITIES	LOADS INLAND VEHICLE	SHIPPING DOCUMENTS SELECT & LOAD SHIP	PAYS FREIGHT	OBTAINS INSURANCE	ASSUMES RISK DURING TRANSIT	PAYS DUTY
MAXIMUM BUYER ↑	EX WORKS	ORIGIN SPECIFIED AS TO PLANT SHIPPING DOCK.	B	B	B	B	B	B
	FOR/FOT	SELLER ARRANGES RAIL CARRIER. OBTAINS BILL OF LADING.	S	—	B	B	B	B
	FAS VESSEL FOREIGN PORT	SAME AS FOB EXCEPT BUYER PAYS FOR LIFTING.	S	B	B	B	B	B
RESPONSIBILITY	FOB VESSEL	SELLER ARRANGES INLAND SHIPPING TO SHIP DOCK.	S	S	B	B	B	B
	C & F	SELLER'S PRICE INCLUDES TRANSPORTATION.	S	S	S	B	B	B
	C I F	SAME AS C & F AND ALSO INCLUDES INSURANCE.	S	S	S	S	B	B
	EX SHIP	SELLER TO IMPORT SHIP LOADING.	S	S	S	S	S	B
	EX QUAY (DOCK)	SELLER PAYS TO IMPORT CUSTOMS.	S	S	S	S	S	S
MAXIMUM SELLER	FOB DELIVERED	SELLER PAYS ALL COST.	S	S	S	S	S	S

POOLER & ASSOCIATES

FIGURE 3-2.   Buyer and Seller Responsibilities.

of price. The "Spuds R' US" case study illustrates a typical competitive purchase and the application of various price analytical techniques to the situation. It also illustrates the possible shortcomings of price analysis. The "Sprint Aircraft" case study illustrates a situation where the buyer may not want to rely solely on catalog pricing to determine whether to accept a catalog-based offer for specialty aircraft.

Janitorial Company

Ms. Katherine Kul, a buyer from the Night Company, issued a sole-source solicitation for 12 months of janitorial services. In response, a proposed firm-fixed-price offer of $120,000 was received from the Janitorial Company. In checking the contracting records, Ms. Kul discovers that a contract was recently awarded by another division of the Night Company for the same type and level of service for a similar size and type of facility for a six-month period at $60,000 ($10,000 per month). The contract by the other division was awarded following competitive bidding. However, only the Janitorial Company responded to the solicitation and the award by the other division was made based on a favorable comparison of the offer with the in-house estimate. Ms. Kul wonders whether price analysis, using the price awarded in the earlier purchase as a basis for comparison, is an effective technique here.

Discussion

Ms. Kul has good reason to be skeptical. The mere fact that a sister division awarded a similar contract for the same monthly price may or may not be reason to accept the current $120,000 offer. The fact the previous contract relied on the in-house estimate should not give Ms. Kul a "warm and fuzzy" feeling. In-house estimates are generally not the most reliable basis for comparison, particularly where the service has never been on contract before or (worse still) where the services were previously performed in-house and those in-house costs are used as a basis for the contract estimate. In addition, there may be a considerable time difference between the award of the previous contract and the current solicitation, making it necessary to adjust the previous contract price for the ravages of inflation or deflation. Lastly, although the services being ordered and facilities being serviced appear "comparable," the probability of their being comparable to the point there is no cost difference between the two is very low. Ms. Kul may want to explore the possibility of getting some cost data from the Janitorial Company that can be used to compare against the current and previous in-house estimate.

Spuds R'Us

In September 1992, the following proposals were received and found technically acceptable in response to a solicitation from the TOT Corporation:

Company	Price	Delivery
Company A	$29.75 each	delivery within 150 days after receipt of order.
Company B	$21.20 each	delivery within 180 days after receipt of order.
Company C	$32.00 each	delivery within 180 days after receipt of order.
Company D	$23.75 each	delivery within 90 days after receipt of order.
Company E	$31.00 each	delivery within 180 days after receipt of order.
Company F	$30.00 each	delivery within 180 days after receipt of order
Company G	$30.50 each	delivery within 150 days after receipt of order

The solicitation had requested the submission of proposals for the manufacture and delivery of 4,000 Spuds. Spuds are essential expendable components of a Tot production line. The purchase request had cited funds in the amount of $110,000 and had requested a required delivery date of 180 days

56   Cost/Price Analysis: Tools to Improve Profit Margins

after receipt of order, with a desired delivery date of 180 days after receipt of order. Since this type item has been competitively purchased in the past, and since the current purchase request contained only a minor modification in the specification of the Spuds, it was determined that a fair and reasonable price could be negotiated on the basis of adequate price competition; therefore price analysis, as opposed to cost analysis, can be used.

The historical files indicated previous purchases of Spuds as follows:

November 1991	2,000 each @ $27.50/ea. from Company B—150 day delivery
July 1990	3,000 each @ $25.00/ea. from Company A—180 day delivery
October 1989	2,500 each @ $19.80/ea. from Company D—180 day delivery

The buyer, Ms. Tatum, is attempting to determine the following:

- What methods of price analysis are appropriate to this purchase.
- The method of utilization of these different methods of analysis.
- Whether an award should be made based on price analysis alone.

Discussion
The following price comparison techniques would be available to Ms. Tatum in her quest to determine a fair and reasonable price:

- Comparison with the in-house estimate.
- Comparison with past prices paid for the same or similar items.
- Comparison with other offers on the instant purchase.

Ms. Tatum knows the purchase request cited funds in the amount of $110,000, which amounts to a unit price of $27.50. One of the primary criteria in determining the usefulness of an in-house estimate for comparison purposes is its validity. Ms. Tatum must ask herself how valid the estimate is. Ms. Tatum would probably conclude the estimate is little more than an indication of the amount available for the purchase, particularly since it coincides with the last purchase price paid in November of 1991. This form of price comparison should not be used unless it can be determined that the estimate resulted from a careful study of the plans, specifications and drawings for the current purchase transaction. This does not appear to be the case here, thus comparison with the in-house estimate is probably not going to help much.

Several factors need to be carefully examined to compare this purchase with past prices paid for the same or similar items. First, if there were any change made to the specifications since the previous purchases, it would be necessary to quantify the cost effect of those modifications. This type of analysis would attempt to determine if the old price, plus the estimated cost of the modification equals, or comes close to, the new price. Company D is offering a price considerably out of line with the others. Care should be taken to assure that Company D understands the requirements of the modification to the specifications.

A second factor to consider in comparing with previous prices is the change in quantity. Normally, it would be expected that an increased quantity would reduce the unit price through the economies of scale. This is not always the case, however. An exception to the rule could pertain when the increased quantity exceeds the supplier's current capacity, requiring the supplier to expand facilities or acquire additional personnel. Before the price comparison technique could be used, a determination of the effect of the increased quantity must be made, and, if possible, quantified.

A third factor to consider in comparing with previous prices is the delivery requirements. Even though they have not changed, some attention should be given to the fact that Company D has offered a considerably accelerated delivery schedule for this purchase, even though it required the full 180-day delivery period when it last won an award. This difference should be investigated, particularly in relation to the following questions:

- Has Company D so improved its production capability that it can expedite delivery?
- Does Company D already have quantities on hand, so it can issue from stock?
- If Company D has Spuds in stock, can it make the required modification and still deliver on time?
- Has Company D so improved its production efficiency that it can delivery so much faster than the others? What is the probable effect on price as a result of this increased efficiency?

A fourth factor to consider in comparing with previous prices are economic changes since the previous purchases. As inflation continues (even though at a reduced rate), it can be expected that the current prices will increase over historical prices. The amount of such an expected increase would have to be quantified before valid comparisons can be made.

It is necessary to ensure that competitive offers were submitted in response to the request for quotations. Adequate price competition exists when there are two or more responsive and responsible offerors independently contending for an award based on the lowest price. There are seven responsive and responsible offerors independently contending for the Spuds contract. Does this automatically mean there has been adequate competition? Before answering this, Ms. Tatum should determine if one of the offerors has such a determinative competitive advantage that it is immune from the effects of competition. In other words, one of the offerors might be able to set its price without considering the competition because it has an advantage (proprietary production process or whatever) that allows it to undersell the competition and still make a profit. The price offered by Company D is so far out of line from the other prices offered that it cannot be considered fair and reasonable based on adequate price competition. The most probable reasons for the price difference are as follows:

- Company D has made a mistake in estimating its price.
- Company D has some kind of competitive advantage.
- Company D wants to "buy in" on the contract. It might do this for any number of reasons, including a willingness to accept a loss in order to keep its production line busy. Another possibility could be that Company D wants the contract to get its "foot in the door," or that the specifications will change enough to recover its losses.

Whether an award should be made based on price analysis alone depends on the answers Ms. Tatum has obtained to all these nagging questions. Ms. Tatum would probably conclude that more information is needed to account for the price offered by Company D. If the differences between its current offer and the last contract can be accounted for and quantified, and if the price on the last contract had been affirmatively determined to be fair and reasonable, it could be possible to determine the current price to be fair and reasonable based on historical comparison. Normally, where the only applicable price comparison method is historical comparison, and where the differences require additional information, a valid determination that the price is fair and reasonable cannot be made without submitting a cost proposal and using cost analysis. Even a limited cost analysis might be helpful in this situation.

Sprint Aircraft

Ms. Sally Port, buyer for World-Wide Widgets, received a purchase request for nine Sprint Model 360 executive aircraft. The president and eight vice-presidents were scheduled to attend a sales meeting in Uzbekistan at the beginning of the following week, and they wanted to impress their potential Uzbeki buyers. They were particularly impressed with the Sprint Model 360 since the company had these existing Sprints, which had been used to real advantage in winning contracts in Nepal and Timbuktu.

The Model 360 is a standard commercial aircraft fabricated and assembled by the Sprint Aircraft Company of Snohomish, Washington (near Seattle). The Model 360 is produced and sold at a rate of 300 to 500 per month.

One year ago Ms. Port awarded Sprint a firm-fixed-price contract for three aircraft at a unit price of $8,200. This award was the result of two-step competitive bidding under which Sprint bid low enough to be reasonably assured it would get the contract.

In response to Ms. Sally Port's solicitation, the Sprint Company submitted the proposal shown in Table 3-9.

Sprint did not submit a cost proposal with its offer because this data was not requested. Sprint established the price of $10,500 per unit by applying a 30 percent discount to the published catalog price of $15,000. The distributor price for a Model 360 is $10,950, which is a 27 percent discount from the catalog price. The Sprint representative stated that 27 percent is the maximum discount to distributors and that no other concessions are made at any time.

Ms. Port concluded that the price met all the requirements for negotiation on price analysis alone, since there is a discount from an established catalog price for the item. Despite this, Ms. Port believed the unit price of $10,500 was unreasonably high considering the estimated cost of the item plus a reasonable profit. She concluded that Sprint's unit price on the last purchase was a break-even price. She estimated that production costs for the aircraft had increased no more than 5 percent since the initial purchase. By making an allowance for a reasonable profit of 10 percent, she could make an estimate of a reasonable price. By her analysis and knowledge of manufacturing costs of this item, she estimated that if the purchase were competitive, comparable aircraft could be purchased at approximately $9,500 per unit.

What courses of action are open to Ms. Port?

What is the estimated unit price of this aircraft using the estimated costs and profit factor of Ms. Port?

Discussion

Ms. Port knows that competition is the key to guaranteeing a fair and reasonable price. Unfortunately, that option is not available to Ms. Port insomuch as her top management is interested specifically in the Sprint Model 360.

Ms. Port could ask Sprint for a cost breakdown of the $10,500 unit price. Chances are, however, that Sprint will decline to provide that information,

TABLE 3-9

Item	Unit Price	Total Price
Model 360	$10,500	$94,500

insomuch as Sprint probably understands the World-Wide Widget management's preference for its Model 360. Ms. Port should ask for this information, recognizing that she probably won't get it.

Without cost information, Ms. Port is left with her ability to build up a price from the knowledge she has gained about the company and the industry. Assuming the labor component of the aircraft is 50 percent of the prior contract's cost of $128,400, she can escalate the labor portion ($64,200) of total estimated cost by 3 percent in order to arrive at an escalated cost of $66,126 for labor. The other-than labor costs of $64,200, when escalated by 7 percent (the increase in the overall Consumer Price Index for the year), becomes $68,694. The total estimated costs would be $66,126 + $68,694 = $134,820. By applying a 5 percent profit, the estimated price is $141,561. This would be a reasonable price objective if Sprint is willing to negotiate.

SUMMARY

The conclusion that a price is fair and reasonable must be based on some form of price or cost analysis. How detailed this analysis will be depends on the dollar value and the nature of the product or service being purchased.

Price analysis is generally used without cost analysis for low-dollar purchases; for most competitive purchases, even though of a large dollar value; for purchases based on existing catalog or market prices; and for purchases of items or services for which regulated prices exist (regulated utility services).

To perform price analysis, the buyer must have a base or reference to which the quoted price can be compared. That basis for comparison must itself be reasonable. Then the buyer must ensure that the quotation and the base are truly comparable. That is why the comparison of competitive quotations is such an effective method of price analysis. The buyer can be reasonably assured that the items are comparable and, presuming that the firms involved are really competing with one another, that the lowest price submitted will be reasonable.

In performing any price analysis, the buyer must consider the quality of the items for which prices are being compared, the quantities involved in the sale, delivery conditions (F.O.B. origin versus F.O.B. destination), and market conditions. If the buyer can arrive at a reasonable base for comparison, even though it includes adjustments for differences in relation to some of these items, price analysis techniques will determine whether a price is fair and reasonable.

Many methods of price analysis are available. Selecting the proper method depends on the specific features of the acquisition situation. In many in-

Price Comparison Methods and How to Use Them 61

stances, a combination of methods is best. The following methods are among the most commonly used:

- Comparison with other prices and quotations submitted.
- Comparison with published catalog or market prices.
- Comparison with prices set by law or regulation.
- Comparison with prices for the same or similar items.
- Comparison with prior quotations for the same or similar items.
- Comparison with market data (indexes).
- Application of rough yardsticks (such as dollars per pound or per horsepower or other units) to highlight significant inconsistencies that warrant additional pricing inquiry.
- Comparison with independent estimates of cost developed by knowledgeable personnel within the buying organization.
- Use of value and/or visual analysis.

4
Elements of Cost

DEFINITIONS: DIRECT AND INDIRECT COSTS

Direct costs are those operational or production-related costs that can be traced to individual units of output or performance of the organization. They are often referred to as traceable, specific, or separable costs. In the construction business, they are referred to as "brick and mortar costs" because they are considered to be an integral part of the finished facility. Direct costs include direct labor, direct material, and other directs.

Indirect costs are those generalized costs that cannot (without a great deal of difficulty) be traced to individual units of output. They are often referred to as nontraceable, common, general, or joint costs. In the construction business, they are referred to as "costs of doing business" because they are considered overhead or general business-related costs. Overhead is often considered synonymous with "burden," particularly in manufacturing environments. Indirect costs may include many different categories and types, including facility costs like maintenance and repair costs; utility service costs, including heat and light; and depreciation on buildings and equipment. Other indirect costs may include manufacturing/operational support labor, fringe benefits, and MRO items. Salaries of the organization's top management, research and development costs, and costs of preparing bids and offers for work are almost always considered indirect. An important point to remember here is that every organization establishes its own chart of accounts to handle its specific situation. As long as that classification system does not violate the law or acceptable financial and/or managerial accounting practice, no one can question that classification system. Another way of saying this is that Company X may classify certain costs as indirect while Company Y may classify similar costs as direct. It is dangerous to generalize and say certain categories of cost are always direct or indirect.

These two cost classifications are important to suppliers in pricing products and services. In the long run, supplies must cover all costs, direct and indirect, to remain solvent. Ideally, suppliers will want to cover these costs and then some to earn a profit for the period in question. In the short run, suppliers may need to price certain products and services at less than full cost recovery to meet the competition or position their products and services at appropriate pricing points. (This concept is where the term "loss leader" comes from, a concept used by grocers to price certain items below cost in order to get more traffic into the store where, hopefully, the consumer will be tempted to buy not only the loss leader but also profit-returning items. In order to price products and services, suppliers will need to know what the direct costs are for all products and services and have a reasonable mechanism to allocate indirect costs to these products and services. Strange things can happen (like the infamous overpriced toilet seat and equally infamous overpriced hammer) when too much indirect cost is allocated to certain items or services.

DEFINITIONS: VARIABLE AND FIXED COSTS

The focus in this section is, for the most part, on historical cost for financial reporting purposes. Another method of classifying cost is necessary for estimating the costs of future operations and performance so that decisions can be made about future pricing of products and services and about how to bid on future work. The emphasis here is on what is called "cost behavior" or "cost-volume-profit" relationships. If the volume of production for item X (or for the organization as a whole) increases (or decreases) to a major degree, what will the impact of that change be on the cost of the supplier's products and services and, more importantly for the supplier, what will the impact of that change be on the profit earned from sales of the products and services?

In order to answer these questions, the supplier must attempt to classify direct and indirect costs throughout the possible ranges of sales and production. By going through this classification process, the supplier will find that certain costs will change directly with production volume, while their unit costs will remain somewhat constant throughout various ranges of production volume. These costs will be categorized as "variable" because they vary with volume. Variable costs generally include labor and materials.

A graphic representation of variable labor costs would look like that shown in Figure 4-1.

The supplier will find that other costs will change little or not at all with production volume, while their unit costs will vary somewhat directly with volume (generally in a decreasing manner as volume increases). These costs will be categorized as "fixed" because they do not vary with volume. Fixed costs generally include costs associated with building rental or lease and other facilities-related costs.

A graphic representation of fixed costs would look like that shown in Figure 4-2.

The supplier will find that other costs will fit neither category, but will instead vary somewhat with volume, albeit not as directly as variable costs. These costs will generally contain a fixed component and a variable component, and are called appropriately "semivariable" or "semifixed." Typically, costs of maintenance, heat, light, power, and other utilities fit this in-between situation.

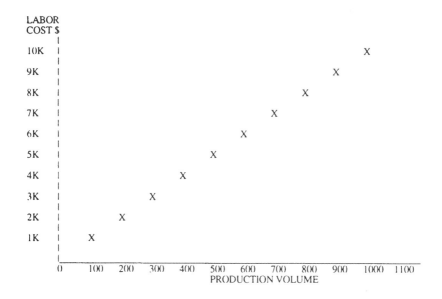

FIGURE 4-1.

66 Cost/Price Analysis: Tools to Improve Profit Margins

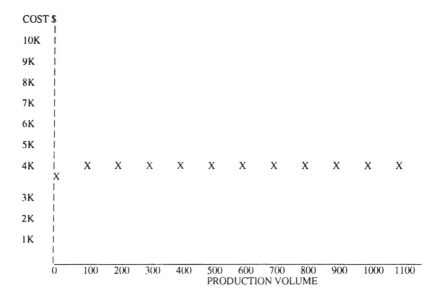

FIGURE 4-2.

A graphic representation of semifixed or semivariable costs would look like that shown in Figure 4-3.

Although indirect costs may vary directly with volume, they are more commonly found to be fixed or semivariable/semifixed. In dealing with these categories of cost, it is appropriate to segment the entire range of production possibilities into volume ranges. A variable cost in one production range may become semivariable at a higher production volume (or vice versa).

The American Company case study demonstrates variable- and fixed-cost behavior in a total and unit sense.

The American Company

Mr. Walks with Pride of the Sioux Nation has a requirement for several hundred miniature canoes. He knows he can only get them from The American Company. This is not all bad, since he knows the production history of The American Company, as it relates to production volume and total production cost. His supervisor, Ms. Leads with Grace, wants Mr. Pride to determine the variable and fixed cost per unit at different levels of production so that she can determine whether American is showing any learning in its production process and if it is managing its fixed costs to her satisfaction.

Mr. Pride has obtained the figures in Table 4-1.

Mr. Pride wants to calculate the information shown in Table 4-2.

Elements of Cost   67

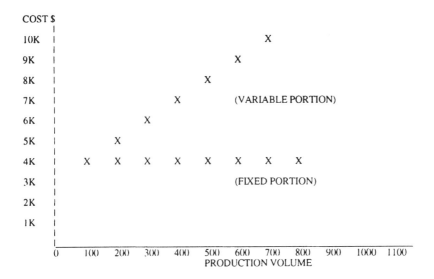

FIGURE 4-3.

Discussion

Mr. Pride will need to approach this in a methodical way. The best approach is to use the given data to derive the next statistic, and so on. Given the volume of production and the total cost, he can derive the total cost per unit by dividing the total cost by the production volume.

This results in the figures shown in Table 4-3.

Given the total cost per unit, he can derive the variable cost per unit by dividing the difference in total cost for two levels of production by the difference in production volume for those same two levels. For the 1,500 and 2,500 levels of production, this would be ($2,250 − $1,750)/(2,500 − 1,500) = $500/1,000 = $.50. Knowing that the variable cost per unit is not likely to change through this production range, Mr. Pride can fill in the variable cost per unit blocks as shown in Table 4-4.

Given the variable cost per unit, Mr. Pride can readily compute the total

TABLE 4-1

Volume of Production (Units)	Total Cost $
1,500	$1,750
2,500	$2,250
3,000	$2,500

68 Cost/Price Analysis: Tools to Improve Profit Margins

TABLE 4-2

Volume of Prod.	Total Cost	Total Cost per Unit	Variable Cost per Unit	Total Variable Cost	Total Fixed Cost	Fixed Cost per Unit
1,500	$1,750	___	___	___	___	___
2,000	___	___	___	___	___	___
2,500	$2,250	___	___	___	___	___
3,000	$2,500	___	___	___	___	___

TABLE 4-3

Volume of Prod.	Total Cost	Total Cost per Unit	Variable Cost per Unit	Total Variable Cost	Total Fixed Cost	Fixed Cost per Unit
1,500	$1,750	$1.17	___	___	___	___
2,000	___	___	___	___	___	___
2,500	$2,250	$0.90	___	___	___	___
3,000	$2,500	$0.83	___	___	___	___

TABLE 4-4

Volume of Prod.	Total Cost	Total Cost per Unit	Variable Cost per Unit	Total Variable Cost	Total Fixed Cost	Fixed Cost per Unit
1,500	$1,750	$1.17	$0.50	___	___	___
2,000	___	___	$0.50	___	___	___
2,500	$2,250	$0.90	$0.50	___	___	___
3,000	$2,500	$0.83	$0.50	___	___	___

variable cost by multiplying the variable cost per unit times the volume of production. He can fill in the total variable cost blocks as shown in Table 4-5.

Given the total variable cost, Mr. Pride can readily compute the total fixed cost by subtracting the total variable cost from the total fixed cost. Knowing that the total fixed cost per unit is not likely to change through this production range, he can fill in all the total fixed cost blocks as shown in Table 4-6.

Elements of Cost   69

TABLE 4-5

Volume of Prod.	Total Cost	Total Cost per Unit	Variable Cost per Unit	Total Variable Cost	Total Fixed Cost	Fixed Cost per Unit
1,500	$1,750	$1.17	$0.50	$7.50	___	___
2,000	___	___	$0.50	$1,000	___	___
2,500	$2,250	$0.90	$0.50	$1,250	___	___
3,000	$2,500	$0.83	$0.50	$1,500	___	___

TABLE 4-6

Volume of Prod.	Total Cost	Total Cost per Unit	Variable Cost per Unit	Total Variable Cost	Total Fixed Cost	Fixed Cost per Unit
1,500	$1,750	$1.17	$0.50	$750	$1,000	___
2,000	___	___	$0.50	$1,000	$1,000	___
2,500	$2,250	$0.90	$0.50	$1,250	$1,000	___
3,000	$2,500	$0.83	$0.50	$1,500	$1,000	___

Given the total variable cost and the total fixed cost for 2,000 units, Mr. Pride can now "backfill" the total cost and total cost per unit for 2,000 units as shown in Table 4-7.

Given the total fixed cost he can derive the fixed cost per unit by dividing the total fixed cost by the production volume. This results in the figures shown in Table 4-8.

These statistics demonstrate that while variable costs increase in a total sense over increases in production volume, variable cost per unit remains

TABLE 4-7

Volume of Prod.	Total Cost	Total Cost per Unit	Variable Cost per Unit	Total Variable Cost	Total Fixed Cost	Fixed Cost per Unit
1,500	$1,750	$1.17	$0.50	$750	$1,000	___
2,000	$2,000	$1.00	$0.50	$1,000	$1,000	___
2,500	$2,250	$0.90	$0.50	$1,250	$1,000	___
3,000	$2,500	$0.83	$0.50	$1,500	$1,000	___

TABLE 4-8

Volume of Prod.	Total Cost	Total Cost per Unit	Variable Cost per Unit	Total Variable Cost	Total Fixed Cost	Fixed Cost per Unit
1,500	$1,750	$1.17	$0.50	$750	$1,000	$0.67
2,000	$2,000	$1.00	$0.50	$1,000	$1,000	$0.50
2,500	$2,250	$0.90	$0.50	$1,250	$1,000	$0.40
3,000	$2,500	$0.83	$0.50	$1,500	$1,000	$0.33

constant. The opposite is true for fixed costs. They remain constant over increases in production volume but decrease in unit cost over that same volume.

Assume American buys a new machine, which results in a 50 percent decline in variable cost per unit and a $500 increase in fixed costs. Compute for the categories shown in Table 4-9.

A decrease in the unit variable cost by 50 percent and an increase in the total fixed costs by $500 results in the figures shown in Table 4-10.

Fixed cost per unit is easily derived by dividing the total fixed cost by the production volume. This results in the figures shown in Table 4-11.

Total variable cost would, of course, be the variable cost per unit times the production volume, as shown in Table 4-12.

Total cost would then be determined by adding total variable cost and total fixed cost. Total cost per unit would then be computed by dividing total cost by volume of production. The results are as shown in Table 4-13.

This computation demonstrates the sharp effects of leverage on costs of production. In this case, we observe that capital equipment (leverage) begins to favorably influence the total unit production cost when production ex-

TABLE 4-9

Volume of Prod.	Total Cost	Total Cost per Unit	Variable Cost per Unit	Total Variable Cost	Total Fixed Cost	Fixed Cost per Unit
1,500	——	——	——	——	——	——
2,000	——	——	——	——	——	——
4,000	——	——	——	——	——	——
8,000	——	——	——	——	——	——
12,000	——	——	——	——	——	——

Elements of Cost 71

TABLE 4-10

Volume of Prod.	Total Cost	Total Cost per Unit	Variable Cost per Unit	Total Variable Cost	Total Fixed Cost	Fixed Cost per Unit
1,500	——	——	$0.25	——	$1,500	——
2,000	——	——	$0.25	——	$1,500	——
4,000	——	——	$0.25	——	$1,500	——
8,000	——	——	$0.25	——	$1,500	——
12,000	——	——	$0.25	——	$1,500	——

TABLE 4-11

Volume of Prod.	Total Cost	Total Cost per Unit	Variable Cost per Unit	Total Variable Cost	Total Fixed Cost	Fixed Cost per Unit
1,500	——	——	$0.25	——	$1,500	$1.00
2,000	——	——	$0.25	——	$1,500	$0.75
4,000	——	——	$0.25	——	$1,500	$0.38
8,000	——	——	$0.25	——	$1,500	$0.19
12,000	——	——	$0.25	——	$1,500	$0.13

TABLE 4-12

Volume of Prod.	Total Cost	Total Cost per Unit	Variable Cost per Unit	Total Variable Cost	Total Fixed Cost	Fixed Cost per Unit
1,500	——	——	$0.25	$ 375	$1,500	$1.00
2,000	——	——	$0.25	$ 500	$1,500	$0.75
4,000	——	——	$0.25	$1,000	$1,500	$0.38
8,000	——	——	$0.25	$2,000	$1,500	$0.19
12,000	——	——	$0.25	$3,000	$1,500	$0.13

TABLE 4-13

Volume of Prod.	Total Cost	Total Cost per Unit	Variable Cost per Unit	Total Variable Cost	Total Fixed Cost	Fixed Cost per Unit
1,500	$1,875	$1.25	$0.25	$ 375	$1,500	$1.00
2,000	$2,000	$1.00	$0.25	$ 500	$1,500	$0.75
4,000	$2,500	$0.63	$0.25	$1,000	$1,500	$0.38
8,000	$3,500	$0.44	$0.25	$2,000	$1,500	$0.19
12,000	$4,500	$0.38	$0.25	$3,000	$1,500	$0.13

ceeds 2,000 units. Without the new equipment, the unit cost for 4,000 would be $0.50 + $1,000/4,000 = $0.75. With the new equipment, the unit cost for 4,000 is $0.63. The differences in favor of the new equipment increase significantly beyond the 4,000 production point.

USING VARIABLE AND FIXED COST TO COMPUTE BREAK-EVEN VOLUME OR SALES PRICE

In addition to using cost behavior as a tool to price products and services and bid on future work, the supplier can use the tool to conduct "break-even" analysis, either for the organization as a whole or for individual products and services. Break-even attempts to answer the question: Given, a planned selling price for products/services and given fixed and variable costs of production, what sales volume is required to cover fixed and variable costs and walk away with no profit/no loss? The algebraic equation that must be solved in this instance is $S \times SV = VC + FC$, where S is Selling Price; SV is Sales Volume; VC is Total Variable Cost (further determined by multiplying Variable Cost Per Unit times Sales Volume); and FC is Total Fixed Cost. Solving for SV, the formula becomes: $SV = (VC + FC)/S$.

Using actual data, solve for break-even production volume given $S = \$14$/unit; $VC = \$10$/unit; and $FC = \$500,000$. Substituting these numbers in the formula $S \times SV = VC = FC$, using X as the unknown sales volume, we have $\$14$/unit $\times X = \$10$/unit $\times X + \$500,000$. Solving for X, we have $\$14X - \$10X = \$500,000$; $\$4X = \$500,000$; $X = 125,000$ units.

Another question that can be answered is: Given a planned sales volume and given fixed and variable costs of production, what sales price must be charged at that volume in order to break-even? The algebraic equation that must be solved in this instance is the same as the previous equation:

$S \times SV = VC + FC$, where S is Selling Price; SV is Sales Volume; VC is Total Variable Cost (further determined by multiplying Variable Cost Per Unit times Sales Volume); and FC is Total Fixed Cost). Solving for S, the formula becomes: $S = (VC + FC)/SV$.

Using actual data, solve for break-even production volume given $SV = 250,000$; $VC = \$10$/unit; and $FC = \$500,000$. Substituting these numbers in the formula $S \times SV = VC = FC$, using X as the unknown sales price, we have $X \times 250,000 = \$10$/unit $\times 250,000 + \$500,000$. Solving for X, we have $X = (\$2,500,000 + \$500,000)/250,000$; $X = \$12$/unit.

Although the buyer can't always determine a supplier's break-even point, he or she should understand that the supplier (and the buyer's own organization) employs this technique, and further, if competition is very keen in the industry, the price being quoted by a specific supplier may be close to break-even for that product (or even below break-even at a point where the sales prices is covering the variable costs of the product but not the full cost allocation, including fixed costs). In competitive purchasing, the buyer doesn't worry about the break-even point (other than to wonder how much profit or loss the successful supplier is earning on the sale). In noncompetitive purchasing, the buyer is extremely interested, because the buyer must understand the cost behavior of the supplier in order to negotiate a truly fair and reasonable price.

The "Sanford Radar Company" case study demonstrates how to graph and compute the break-even point for a single product company (the simplest case).

Sanford Radar Company

Larry Mitchell is a buyer for the State of Texas. The State has a requirement for several hundred Police Radars for its Texas Rangers. The State has purchased these radars for the last several years from the Sanford Radar Company. Larry fully intends to purchase the next requirement from Sanford, but he has been hearing bad things about the company. He understands from talking to Fred, the Sanford marketing representative, that the company may be in dire straits financially because it has had to sell quite a number of its radars to law enforcement agencies at a low markup or even at break-even (or below) to meet the fierce competition. Larry thinks he can obtain Sanford's production cost data for different levels of production because he has been such a good customer through the years. Moreover, if Sanford is indeed in trouble, they may be eager to provide this data in order to substantiate a reasonable profit on the sale. In response to Larry's solicitation request for production cost data, Sanford provided the information shown in Table 4-14.

74   Cost/Price Analysis: Tools to Improve Profit Margins

TABLE 4-14

Production	Total Production Cost
500 Units	$1,125,000
1000 Units	$1,500,000
2000 Units	$2,250,000
3000 Units	$3,000,000
4000 Units	$3,750,000

Larry wants to determine the fixed costs and variable costs of production in order to determine the algebraic equation of the total cost line. He wants to use that equation to determine the fixed, variable, and total costs of production at 2,500 units and at 2,712 units. In addition, he wants to compute the break even sales price for Sanford at 2,500 units of production.

Discussion

Larry's plot of the production data appears in Figure 4-4.

Larry knows he can determine the fixed, variable, and total production costa for both 2,500 and 2,712 units by reading the data directly from the graph, but he believes that would not be nearly as accurate as computing the data, given the equation for the line. He knows that he can develop the equation using the formula Y (production cost) $= a$ (the point where the total cost line crosses the Y axis) $+ b$ (the slope of the total cost line) times x (units of production). Another way of explaining this is that the total cost of

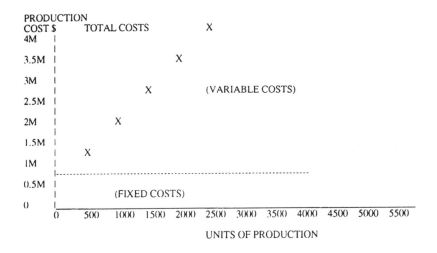

FIGURE 4-4.

production is equal to the fixed cost of production plus the variable cost per unit of production times the quantity produced.

Larry knows that he can compute the slope of the line by taking two points on the total cost line, determining the difference in production cost ($Y2 - Y1$) for these two points and the difference in units of production ($X2 - X1$) for these two points and then dividing the difference in production cost by the difference in units of production. He can readily see that two good total cost points for this would be $1,500,000 and $3,000,000. The corresponding unit production values for these two points would be 1,000 and 3,000, respectively. The difference in production cost would be $3,000,000 − $1,500,000 or $1,500,000. The difference in unit production would be 3,000 − 1,000 or 2,000. Dividing the difference in production cost of $1,500,000 by the difference in unit production of 2,000 would yield a value of $750. This is the slope of the total cost line or the b in the equation $Y = a + bX$.

The computation of the point where the total cost line crosses the Y axis (the a in the equation $Y = a + bX$) can be computed by the following formula: $a = Y1 - bX1$. Using the same values he used before for $Y1$ ($1,500,000) and $X1$ (1,000), and the value for b ($750), which he just computed, he can derive the a as follows: $a = \$1,500,000 - (\$750)(1,000)$ or $a = \$1,500,000 - \$750,000 = \$750,000$. This is the value on the Y axis (production cost) where the total cost line crosses that axis. Larry can easily verify that by looking at his graphic representation of that fact.

Larry now wants to know the fixed cost, variable cost, and total cost for 2,500 units. The fixed cost is easily determinable because it will be the value of Y where the total cost line crosses the Y axis. This is, of course, the a in our equation, or $750,000. To determine variable and total cost, he can use the graph by reading the values that correspond to 2,500 units. He can determine them more accurately by using his formula of $Y = \$750,000 + \$750(X)$, where the $750 ($X$) is the variable cost and Y is the total cost. By substituting 2,500 for X, he can compute the variable and total cost as follows: $Y = \$750,000 + (\$750)(2,500)$ or $Y = \$750,000 + \$1,875,000 = \$2,625,000$.

To do the same computation for 2,712 units, we can see that $750,000 continues to be the fixed cost, while the variable and total cost is available by substitution of 2,712 for the value of X in the formula, as follows: $Y = \$750,000 + \$750(2,712) = \$750,000 + \$2,034,000 = \$2,784,000$.

For Larry to compute the break-even price for 2,500, he will have to set the fixed and variable cost for 2,500 units equal to the sales price per unit times 2,500 units. In other words, his formula would be:

$$S(X) = \$750,000 + \$750\,X), \text{ where } S \text{ is the sales price per unit.}$$
Substituting for X, we have:
$$S(2,500) = \$750,000 + \$750(2,500) = \$750,000 + \$2,625,000.$$

Solving for S, we have

($750,000 + $1,875,000)/2,500 = $2,625,000/2,500 = $1,050.

(To break-even at 2,500 units, Sanford would have to charge $1,050 per unit.) This discussion is important because the buyer in noncompetitive purchasing should negotiate a price that covers all direct and indirect costs and returns a reasonable profit to the supplier. Without knowledge of the supplier's cost behavior, such a task will be exceedingly difficult.

SUMMARY

Direct costs are those operational or production-related costs that can be traced to individual units of output or organizational performance. They are often referred to as traceable, specific, or separable costs. Directs costs include direct labor and direct material.

Indirect costs are those generalized costs that cannot (without a great deal of difficulty) be traced to individual units of output. They are often referred to as nontraceable, common, general, or joint costs. They are often referred to as costs of doing business because they are considered overhead or general business-related costs. Overhead is often considered synonymous with burden, particularly in manufacturing environments. Indirect costs may include many different categories and types, including facility costs like maintenance and repair costs; utility service costs, including heat and light; and depreciation on the buildings and equipment. Other indirect costs may include manufacturing/operational support labor, fringe benefits, and MRO items. The salaries of the organization's top management, research and development costs, and costs or preparing bids and offers for work are almost always considered indirect.

These two classifications of costs are important to the supplier in pricing products and services. In the long run, suppliers must cover all costs, direct and indirect, to remain solvent. Ideally, suppliers will want to cover these costs and then some to earn a profit for the period in question. In the short run, suppliers may need to price certain products and services at less than full cost recovery to meet the competition or to position their products and services at appropriate pricing points. Suppliers therefore need to know what the direct costs are for all products and services and then have a reasonable mechanism to allocate his indirect costs to these products and services.

Another method of classifying cost is necessary for estimating the costs of future operations and performance so that decisions can be made about future pricing of products and services and about how to bid on future work. The emphasis here is on cost behavior or cost-volume-profit relationships. If the volume of production for item X (or for the organization as a whole) increases (or decreases) to a major degree, what will be the impact of that change on the cost of the supplier's products and services and, more importantly for the supplier, what will be impact of that change on the profit earned from sales of the products and services?

The supplier must attempt to classify direct and indirect costs throughout the possible ranges of sales and production. By going through this classification process, the supplier will find that certain costs will change directly with production volume, while unit costs will remain somewhat constant throughout various ranges of production volume. These costs will be categorized as variable because they vary with volume. Variable costs generally include labor and materials.

In addition to using cost behavior as a tool to price products and services and bid on future work, the supplier can use the tool to conduct break-even analysis, either for the organization as a whole or for individual products and services. Break-even attempts to answer the question: Given, a planned selling price for products/services and given fixed and variable costs of production, what sales volume is required to cover fixed and variable costs and walk away with no profit/no loss?

5
Estimating Cost and Obtaining Cost Proposals From Suppliers

WHY ESTIMATING IS IMPORTANT

A knowledge of cost estimating is essential for both buyers and sellers. Without a consistently reliable estimating system, sellers will not be able to price their goods and services to cover all of their costs and return a viable profit. Without a knowledge of cost estimating procedures in general, and an ability to understand specific supplier cost estimating systems, buyers will be unable to assure management that they have negotiated a fair and reasonable price in the goods and services they buy in noncompetitive purchasing transactions.

METHODS OF ESTIMATING

There are three basic methods of estimating. These methods are round-table, comparison, and detailed.

Round-Table

The round-table method of estimating is commonly employed when there is little, if any, prior cost history with the product or service being purchased. In round-table estimating, members of the Engineering, Manufacturing, Contracts, Purchasing, and Accounting departments gather to apply their experience, knowledge of the product or service, and knowledge of market conditions to arrive at an informed "guesstimate" of the cost to produce and sell the product or service. This type of estimate is generally prepared without the benefit of detailed drawings or bills of material and with very limited

80 Cost/Price Analysis: Tools to Improve Profit Margins

information concerning specifications. This type of estimating is fast and relatively inexpensive.

Comparison

The comparison method of estimating is commonly employed in situations where there is prior cost history for similar products or services but no specific cost experience on the product or service being purchased. The estimator (with assistance from Manufacturing or Engineering, as appropriate) selects cost data from parts or processes comparable to those required for the task. He or she then adjusts the known costs of those similar items by adding or subtracting elements of material and time as determined necessary for the new task.

The "Ranger Company" case study illustrates how knowledge of a similar item, combined with use of "formula estimating" can derive an estimate for a manufactured item new to production.

Ranger Company

Ms. Clara Call, cost estimator, knows her company, the Ranger Company, is going to be doing a lot of bidding and proposing on small government and commercial contracts. The Ranger Company does many small job shop production contracts and needs to be able to quote quickly and accurately to its various customers. Ms. Call wants to have some means of rapidly estimating the cost of various jobs, given only the number of hours of direct labor for the job, the value of material to be used on the job, the hourly wage of the skilled worker doing the job, and the overhead to be charged to the job. All of these Ms. Call obtains from her engineer, Bill Kieth. Ms. Call knows from historical experience that her tool rehabilitation costs are 2 percent of direct labor hours, set up costs are 10 percent of direct labor hours, rework costs are 3 percent of direct labor hours, and nonproduction time is 1 percent of direct labor hours. She further knows that her material price variance is 2 percent of material cost, test costs are 1 percent of material cost, and material burden is 3 percent of material cost. Ms. Call gets a quotation request from one of her commercial customers. Mr. Kieth tells Ms. Call that the job will require 20 hours of labor and $200 of material, that the worker to do the job is paid $25.00 per hour, and that the overhead will be $75.00 per hour. Ms. Call normally uses a factor of 15 percent to cover her general and administrative expenses and profit requirements. Ms. Call wants to determine the amount to quote on this job.

Discussion

Ms. Call should establish her cost estimating formula first, then plug the numbers into the formula. Based on this information, Ms. Call formula would be as shown in Table 5-1.

Using this data, we can compute the value of the job as follows:

Labor and Overhead Cost = 20 (Estimated Direct Labor Hours) × $100.00 (Combined Hourly Labor and Overhead Rate) × 1.16 (Total Labor Factor expressed as a decimal) = $2,320

Material Cost = $200 (Basic Material Cost) × 1.06 (Total Material Factor expressed as a decimal) = $212

The value of the job would then be the combination of these costs plus 15 percent (which Ms. Call believes is needed to cover G&A and profit) or:

($2,320 + $212) × 1.15 = $2,911.80.

Ms. Call decides she will round down (to $2,900) to make sure she gets the job.

Detailed

The detailed method of estimating is commonly employed when there is prior cost history for the product or service being purchased. It is characterized by a thorough, detailed analysis of all work phases, components, processes, and assemblies. Requirements for labor, tooling, material, and additional capital items are produced by this type of estimating. The application of labor rates, material prices, and overhead to the calculated requirements translates the

TABLE 5-1

Cost Element	Percent Factor
Estimated direct labor hours	100
Tool rehabilitation	2
Setup	10
Rework	3
Nonproduction time	1
Total labor factor	116
Basic material cost	100
Price variance	2
Test	1
Material burden	3
Total material factor	106

estimate into dollars. The method is characterized by the presence of complete calculations, records, and quotations that are available for future use. To perform this method of estimating, each component is separated into parts, operations, and cost elements. Data normally developed from this process includes drawings, bills of material, specifications, production quantities, production rates, analysis of manufacturing processes, tooling and capital costs, machine and workstation work loads, plant layout, labor, raw materials and purchased parts, overhead, special tools and dies; manufacturing, engineering, and tooling labor; labor efficiency, setup, and rework; and material scrap, waste, and spoilage. The cost estimator using this method normally relies on the Engineering and Manufacturing departments for estimates of direct labor hours and special tooling costs, the Purchasing department for estimates of material and subcontract costs; the Project Office for estimates of direct travel and direct labor hours; Drafting or Data Control for estimates of costs of data items; and Facility Engineering for estimates of facilities costs. With this data in hand, the cost estimator obtains labor rates from payroll records, overhead rates from accounting, and inflation assumptions and profit rates from management. An alternative method would be to have each department involved in the work submit their own direct cost estimates to the cost estimator. The estimator would consolidate the various departmental estimates and apply overhead and profit factors.

DETAILED ESTIMATING EXAMPLE

An example of how to prepare an itemized and priced bill of materials (necessary for detailed estimating) is found in the "Off-Line Construction" case study. Examples of estimating forms commonly used by both buyers and suppliers in estimating architect engineering and engineering services, construction services, and nonpersonal services are shown in Figures 5-1, 5-2, and 5-3.

Off-Line Construction

Ms. Amanda Kerr is a buyer for the ABC Corporation. She has a requirement for 1,000 galvanized steel boxes to be built to specification. The requestor provided a unit material cost estimate of $175 per box, based on his recollection of the price paid ($300 per box) to Off-Line Construction on the last purchase. Ms. Kerr is not happy with this estimate, and has asked her construction cost estimator, Mr. Henry Kent, to prepare a detailed, bottom-up estimate of cost for the boxes.

Discussion
Based on his 20 years of experience, Mr. Kent believes he can develop a detailed materials estimate using the following steps: (1) obtain drawings and specifications; (2) list all materials indicated on drawings and specification; (3) calculate the quantity required of each material (including scrap); (4) determine prices of materials from vendors, catalogs, and so forth; (5) extend the quantities of materials by their unit prices (monetize the quantities); and (6) adjust for rejects and rework.

Steps 1 and 2. Obtain Drawings and specifications, and list all materials indicated on drawings and specification.

From Mr. Kent's review of the specifications and drawing, he can calculate the types and amounts of each material required per box. His review determines the following materials are needed: galvanized steel, 1 1/2-inch bolts, 1/2-inch bolts, welding material, handles, hasps, and hinges.

Step 3. Calculate the quantity required of each material (including scrap).

Galvanized steel sheets: Galvanized steel sheets of the type needed for the box are normally manufactured in 4' × 8' sheets. Therefore, the first requirement is to determine how much steel is required per box. By carefully examining the total steel requirements from the sheets, Mr. Kent determines that all box sides can be made from one sheet of steel.

1 1/2 inch bolts: To determine the number of 1 1/2-inch bolts needed, Mr. Kent calculates the following:

Ends:	4 rows × 23 inches/row × bolts/4 inches =	24
Bottom:	2 rows × 24 inches/row × bolts/4 inches =	12
Bottom:	2 rows × 36 inches/row × bolts/4 inches =	18
Credit:	4 bolts at bottom corner since he counted = them twice	− 4
	Total:	50 each 1 1/2-inch bolts

1/2-inch bolts: To determine the number of 1/2-inch bolts, Mr. Kent calculates the following:

Hinges	2 hinges @ 6 each	12
Hasps	7 each	7
Handles	2 handles @ 4 each	8
	TOTAL:	27 each 1/2-inch bolts

Welding material: Mr. Kent estimates each box will require .20 of a pound of welding material.

84 Cost/Price Analysis: Tools to Improve Profit Margins

TABLE 5-2

Material Item Required	Quantity Computation	Quantity
Galvanized steel	1,000 boxes × 1 sheet/box	1,000 sheets 4' × 8" steel
Scrap steel	1,000 × 199 square inches × 0.040 lbs/square inch	7960 pounds of scrap
1 ½" bolts	1,000 × 50 bolts × gross/144 bolts	350 gross
½" bolts	1,000 × 27 × gross/144 bolts	190 gross
Welding material	1,000 × .20 pound	200 pounds
Handles	1,000 × 2	2000 each
Hasps	1,000 × 1	1000 each
Hinges	1,000 × 1 pair	1000 pair

TABLE 5-3

Material Item	Description	Price
Steel	½-inch, galvanized	$89/sheet
Scrap steel	Credit from local scrap collector	$0.75/pound
1 ½" bolts	Steel	$50/gross
½" bolts	Steel	$40/gross
Welding material		$10/pound
Handles	Steel	$10/each
Hasps	Steel	$10/each
Hinges	Steel, strap, 4 inches	$10/pair

TABLE 5-4

Material Item	Quantity Required	Price	Total Cost
Steel	1,000 sheets 4' × 8" GalStl	$89/sheet	$ 89,000.00
Scrap steel	7,960 pounds of scrap	$0.75/pound	($ 5,970.00)
1 ½" bolts	350 gross	$50/gross	$ 17,500.00
½" bolts	190 gross	$40/gross	$ 7,600.00
Welding material	200 pounds	$10/pound	$ 2,000.00
Handles	2,000 each	$10/each	$ 20,000.00
Hasps	1,000 each	$10/each	$ 10,000.00
Hinges	1,000 pair	$10/pair	$ 10,000.00
Total material cost for 1,000 units			$150,130.00
Material cost per unit			$ 150.13

Handles: Mr. Kent determines each box will require two handles.
Hasps: Mr. Kent determines each box will require one hasp.
Hinges: Mr. Kent determines each box will require two hinges.
Scrap: Mr. Kent knows there will be some scrap from the process and calculates he can use all but 199 square inches of the steel on the box.

Based on these estimates, Mr. Kent summarizes the quantities and types of materials as shown in Table 5-2.

Step 4. Determine prices of materials from vendors, catalogs, and so forth. In his next step, Mr. Kent extracts pricing and rate data from catalogs and from vendor quotations as shown in Table 5-3.

Step 5. Extend the quantities of materials by their unit prices (monetize the quantities) (Table 5-4).

Step 6: Adjust for rejects and rework. From past experience, one out of 1,000 boxes is rejected. Thus Off-Line Construction will probably need to build one extra box to make up for the expected reject. One way to adjust the total material cost for the anticipated reject is to add the unit material cost of $150.13 to the total material cost of 1,000 units as follows: $150,130.00 + $150.13 = $150,280.13. To absorb this new total material cost (1,001 units) across the 1,000 units, divide by 1,000 and obtain a new unit material cost as follows:

$$\$150,280.13/1,000 \text{ boxes} = \$150.28 \text{ per box}$$

Combination of Methods

It is important to understand that a given estimate may be based on one or more of these methods. Certain phases of work (in service purchasing) or components (supply purchasing) may lend themselves to detailed estimating while others will need either round-table or comparison estimating. Whichever of the three methods of estimating is used, the supplier's management will have the last say on the estimate, and may indeed adjust the estimate to accommodate management objectives not known to the estimators. A typical example of this would be management's adjustment of the total (upward or downward) in consideration of current or future competition or different contract-type objectives.

ESTIMATING AND ITS RELATIONSHIP TO THE PURCHASING CYCLE

Another very important point to remember is that the buying organization must endeavor to prepare for its own purposes an estimate of cost prior to receiving cost proposals from the supplier(s). In purchasing services, including architect engineering, construction, and nonpersonal services, this estimate should be as detailed as possible, with heavy preference given to using detailed estimating as opposed to round-table or comparison estimating. A detailed in-house estimate is a good first step in the cost analysis process because the quantities of labor, material, and equipment in that estimate often form the basis for the buyer's quantitative positions during the cost analysis of the supplier's cost proposal (estimate) and the negotiation process that follows that analysis.

OBTAINING COST PROPOSALS (ESTIMATES) FROM SUPPLIERS

Buyer Methods and Techniques for Obtaining Cost Data From Suppliers

Before performing a cost analysis, the buyer will need to obtain supplier cost proposal(s) for the needed supplies or services. Firms operating under the "Federal Norm" will be aided in that process by the so-called Truth-in-Negotiations Act, which requires firms doing business with the U.S. government or its contractors to submit "Cost or Pricing Data" on non-competitive purchases and further to certify that data as "Current, Accurate, and Complete" at the time of agreement on contract price. Firms not working under the Federal Norm will be required to use their negotiation skills to get the data they need for price negotiation. Full-blown, detailed cost proposal submissions are the exception rather than the rule in commercial and industrial purchasing. Firms with established, long-term relationships with certain buyers may not be hesitant to provide such full disclosure of data, but even those suppliers will be hesitant to provide the detail generally required under the Federal Norm. The "Non-Federal Norm" will almost always be partial cost data submission and no granting of access to books and records. With that in mind, buyers should attempt to get full disclosure and work with what they get. To aid in this process, buyers should generally attempt to get supplier cost data on the same forms used by their own staff in preparing in-house estimates. Several examples of estimating forms commonly used by both buyers and suppliers in estimating architect/engineering and engineering services, construction services, and nonpersonal services are available. The author recommends using estimating forms similar to those shown in Figures 5-1, 5-2, and 5-3. If the supplies or services have never been bought before, and the buyer is unsure of the cost account structure used in the supplier firm, imposing a cost proposal format on the supplier may not be helpful or appropriate. In those circumstances, it may be necessary to negotiate the cost proposal format.

Estimating Cost and Obtaining Cost Proposals From Suppliers

COST ESTIMATING FORM - ARCHITECT-ENGINEERING SERVICES

DESIGN SERVICES (ONLY) — Sheet 1 of 2 — Estimator:

Contract	Activity	Location	Est Cost of Construction
Project Title			

Labor Category	# of Drawings	Est # of Hours	Hourly Rate	Direct Cost
Project Engineer				
Architect				
Architectural Draftsman				
Structural Engineer				
Structural Draftsman				
Mechanical Engineer				
Mechanical Draftsman				
Electrical Engineer				
Electrical Draftsman				
Civil Engineer				
Civil Draftsman				
Landscape Architect				
Landscape Draftsman				
Spec/Report Writer				
Typist				
Cost Estimator				
Other(s) (Specify)				
TOTAL				

OVERHEAD ___% X $_____ = _____

PROFIT ___% X $_____ = _____

STATE TAX ___% X $_____ = _____

TOTAL DESIGN COST _____

FIGURE 5-1. Cost estimating form for use with architect engineering services—page 1 of 2.

Using Spreadsheet Programs

Another helpful tool used by many buyers is to specify a spreadsheet program to be used for cost proposal submission and to provide the required cost proposal format on disk formatted to that program. This aids the analysis because the buyer can array his or her cost positions alongside the supplier's proposed costs.

88 Cost/Price Analysis: Tools to Improve Profit Margins

COST ESTIMATING FORM - ARCHITECT-ENGINEERING SERVICES
Sheet 2 of 2

ENGINEERING SERVICES (ONLY)   Estimator:

Contract	Activity	Location	Est Cost of Construction
Project Title			

ITEM	# of Drawings	Est # of Hours	Hourly Rate	Direct Cost
FIELD INVESTIGATIONS				
Architect				
Engineers				
APPROVE SHOP DRAWINGS & MATERIALS				
Architect				
Engineers				
AS-BUILT DRAWINGS				
Architect/Engineer				
Draftsman				
INTERIOR DESIGN				
CONSULTATION				
	TOTAL OF ABOVE			
	OVERHEAD ___% X ___ =			
	PROFIT ___% X ___ =			
	STATE TAX ___% X ___ =			
REPRODUCTION	Drawings$	Specifications$	Others$	TOTAL REPRO
SURVEY: ___ Party Days @ _____ Per Party Day =				
SOIL STUDIES: ___ Borings, Depth ___ FT_____ =				
TRAVEL - PER DIEM				
TOTAL ENGINEERING SERVICES				

FIGURE 5-1. (*continued*)   Cost estimating form for use with architect engineering services—page 2 of 2.

Avoid Unnecessary Requests for Cost Data

The buyer should generally avoid asking for unnecessary cost data. When a sole supplier is selling at regulated prices or has catalog prices on its goods and services and an established record of selling from that catalog to all its customers, cost data is unnecessary. When relying on catalog pricing, it is important to understand that suppliers often sell for discounts from catalog

Estimating Cost and Obtaining Cost Proposals From Suppliers 89

CONSTRUCTION COST ESTIMATING FORM - WORKSHEET

Sheet 1 of 1
Identification No.

Contract	Activity	Location	Cost Accounting Code
Project Title			

Item (or Feature) Description Abbreviate if necessary	Quantities		Material Costs		Labor Cost		Engineer Estimate	
	M.H.	No. of Units	Unit	Unit Cost	Cost	Unit Cost	Cost	Unit Cost	Cost

Prepared by (name)	Approved By	Title	Date

FIGURE 5-2. Cost estimating form for use with construction services.

and that any purchase may qualify for one of those discounts. The wise buyer should always obtain whatever discounts are available.

Allow Suppliers Time to Prepare Lengthy Cost Proposals

Generally, the supplier should be provided as much time as the purchase allows to prepare a cost proposal. Supplier cost estimating for larger procure-

ESTIMATE FOR SERVICE WORK

Page 1 of 4

Contract No.	Contract Title	Date

Description of the Work

Estimator

STRAIGHT TIME LABOR

LABOR CATEGORY	HRS	RATE	AMOUNT
1.			
2.			
3.			
4.			
5.			
6.			
7.			
8.			
9.			
10.			
TOTAL (TO BE CARRIED TO LINE 1, PAGE 3)			

OVERTIME

LABOR CATEGORY	HRS	RATE	AMOUNT
1.			
2.			
3.			
4.			
5.			
6.			
7.			
8.			
9.			
10.			
TOTAL (TO BE CARRIED TO LINE 2, PAGE 3)			

REMARKS

FIGURE 5-3. Cost estimating form for use with general service work—page 1 of 4.

ments may take weeks. Preparation time for cost proposal and submission may also become a matter for negotiation if necessary.

SUMMARY

A knowledge of cost estimating is essential for both buyers and sellers. Without a consistently reliable estimating system, sellers will not be able to price their goods and services to cover all of their costs and return a viable profit. Without a knowledge of cost estimating procedures in general, and an ability to understand specific supplier cost estimating systems, buyers will be

ESTIMATE FOR SERVICE WORK

Contract No.	Contract Title	Date
Description of the Work		
Estimator		

MATERIALS AND SUPPLIES

ITEM CATEGORY	QTY	RATE	AMOUNT
1.			
2.			
3.			
4.			
5.			
6.			
7.			
8.			
9.			
10.			
TOTAL (TO BE CARRIED TO LINE 14, PAGE 4)			

EQUIPMENT RENTAL

ITEM CATEGORY	QTY	RATE	AMOUNT
1.			
2.			
3.			
4.			
5.			
6.			
7.			
8.			
9.			
10.			
TOTAL (TO BE CARRIED TO LINE 16, PAGE 4)			

REMARKS

FIGURE 5-3. (*continued*) Cost estimating form for use with general service work—page 2 of 4.

unable to assure management that they have negotiated fair and reasonable prices in the goods and services they buy in noncompetitive purchasing transactions. There are three basic methods of estimating.

The *round-table method* of estimating is commonly employed when there is little, if any, prior cost history with the product or service being purchased. In round-table estimating, members of the Engineering, Manufacturing, Contracts, Purchasing, and Accounting departments gather to apply their experience, knowledge of the product or service, and knowledge of market conditions to arrive at an informed guesstimate of the cost to produce and sell the product or service. This type of estimate is generally prepared without

ESTIMATE FOR SERVICE WORK

Page 3 of 4

Contract No.	Contract Title	Date

Description of the Work

Estimator

SUPPLIER'S WORK	AMOUNT
LABOR AND FRINGE BENEFITS	
1. Labor (Straight Time) (From Page 1)	
2. Labor (Overtime) (From Page 1)	
3. Vacation ____ HRS x ____ BASIC PAY ____ HRS x ____ BASIC PAY ____ HRS x ____ BASIC PAY	
4. Holiday ____ HRS x ____ BASIC PAY ____ HRS x ____ BASIC PAY ____ HRS x ____ BASIC PAY	
5. Subtotal (1 through 4 above)	
6. Health & Welfare ____ HRS x ____ HEALTH & WELFARE	
7. Workmen's Compensation ____ % of #5 above	
8. Payroll Taxes ____ % of #5 above	
9. Other (Pension and Sick Leave)	
10. TOTAL LABOR AND FRINGE COST (1 to 9 above)	
11. FIELD OVERHEAD Supervision ____ HRS x ____ PAY Other (Clerical) ____	

REMARKS

FIGURE 5-3. (*continued*) Cost estimating form for use with general service work—page 3 of 4.

the benefit of detailed drawings or bills of material and with very limited information concerning specifications. This type of estimating is fast and relatively inexpensive.

The *comparison method* of estimating is commonly employed when there is prior cost history for similar products or services, but no specific cost experience on the product or service being purchased. The estimator (with assistance from Manufacturing Engineering, as appropriate) selects cost data from parts or processes comparable to those required for the task. The estimator then adjusts the known costs of those similar items by adding or subtracting elements of material and time as determined necessary for the new task.

Estimating Cost and Obtaining Cost Proposals From Suppliers 93

ESTIMATE FOR SERVICE WORK

Page 4 of 4

Contract No.	Contract Title	Date
Description of the Work		

Estimator		

SUPPLIER'S WORK	AMOUNT
TOTAL COST	
12. Total Labor and Fringe Cost (From Line 10, Page 3)	
13. Liability Insurance ____ % of Line 5, Page 3	
14. Materials and Supplies (From Page 2)	
15. Sales Tax ____ % of Line 14	
16. Rental Equipment (From Page 2)	
17. Rental Sales Tax ____ % of Line 16	
18. Operating & Minor Maint of Owned Equipment	
19. Subtotal, 12 through 18 above	
20. Field Overhead (From Line 11, Page 3)	
21. Subtotal, 19 and 20 above	
22. Home Office Overhead ____ % of Line 21	
23. Equipment Ownership Expense	
24. Subcontractors' Costs	
25. Overhead on Subcontractors' Costs	
26. Subtotal, 21 through 25 above	
PROFIT	
27. Profit ____ % of Line 26	
PRICE BEFORE BOND	
28. Subtotal, 26 and 27 above	
BOND	
29. Bond ____ % of Line 28	
TOTAL PRICE	
30. Total Estimated Price	

REMARKS

FIGURE 5-3. (*continued*)   Cost estimating form for use with general service work—page 4 of 4.

The *detailed method* of estimating is commonly employed when there is prior cost history for the product or service being purchased. It is characterized by a thorough, detailed analysis of all work phases, components, processes, and assemblies. Requirements for labor, tooling, material, and additional capital items are produced by this type of estimating. The application of labor rates, material prices, and overhead to the calculated requirements translates the estimate into dollars. The method is characterized by the presence of complete calculations, records, and quotations that are available for future use. In performing this method of estimating, each component is separated into parts, operations, and cost elements. Data normally developed from this process includes drawings, bills of material, specifications, production quantities, production rates, analysis of manufacturing processes, tooling and capital costs, machine and workstation work loads, plant layout, labor, raw materials and purchased parts, overhead, special tools and dies; manufacturing, engineering, and tooling labor; labor efficiency, setup, rework; and material scrap, waste, and spoilage.

It is important to understand a given estimate may be based on one or more of these methods. Certain phases of work (in service purchasing) or components (supply purchasing) may lend themselves to detailed estimating while others will need either round-table or comparison estimating. Whichever of the three methods of estimating is used, the supplier's management will have the last say on the estimate, and may indeed adjust the estimate to accommodate management objectives not known to the estimators. A typical example of this would be management's adjustment of the total (upward or downward) in consideration of current or future competition or different contract-type objectives.

Before performing cost analysis, buyers will need to obtain supplier cost proposal(s) (supplier estimates) for the needed supplies or services. Buyers should attempt to obtain the level of cost detail they believe necessary to understand their suppliers' cost and pricing methodology. To aid in this process, buyers should generally attempt to get supplier cost data on the same forms used by their own staff in preparing the in-house estimate. Several examples of estimating forms commonly used by both buyers and suppliers in estimating architect-engineering and engineering services, construction services, and nonpersonal services are provided in this chapter. If the supplies or services have never been bought before, and the buyer is unsure of the cost account structure used in the supplier firm, imposing a cost proposal format on the supplier may not be helpful or appropriate. It may be necessary to negotiate the cost proposal format.

Another helpful tool used by many buyers is to specify a spreadsheet program to be used for cost proposal submission and to provide the required cost proposal format on disk formatted to that specified program. This often aids the analysis because the buyer can array his or her cost positions alongside the supplier's proposed costs.

The buyer should generally avoid asking for unnecessary cost data and provide the supplier as much time as the purchase allows to prepare a cost proposal. Supplier cost estimating for larger procurements may take weeks. Preparation time for cost proposal and submission may also become a matter for negotiation if necessary.

6
Performing Cost Analysis, Including Engineering Analysis and Accounting Analysis

INTRODUCTION

The buyer who has requested and received detailed cost proposals will have a considerable amount of work to do. Full-blown cost analysis (supported by price analysis) will generally be required for purchasing items or services using complex buyer specifications; for major noncompetitive purchases; for purchases for which selection is based on factors other than price; and for use of other than firm-fixed-price or fixed-price-with-economic price adjustment (escalation) type contracts.

USING THE PRICING TEAM

In such circumstances (not the normal rule, since most procurements will not require cost proposal submission because of the presence of price competition) the buyer will generally find it necessary to request other functional experts to help conduct engineering and accounting analysis. The dividing line between the two is somewhat easily determined in that engineering analysis devotes itself to a quantitative analysis of labor hours, material quantities, and equipment quantities while accounting analysis devotes itself to a rate analysis (labor rates, materials prices, and overhead rates). The engineering analysis can assist in determining whether the cost proposals reflect an adequate understanding of the work. Evaluating the direct cost elements can reveal whether the proposer has adequately provided for the required services, materials, equipment, travel, and other expenses. If the offeror fails to address areas of work in the cost proposal, the buyer should question whether the offeror really understands the work. This evaluation should address the following:

- The completeness of the supplier's proposed costs. The degree to which the supplier correlates and allocates the labor, material, and other resources to the work plan will directly influence the efficiency, productivity, and schedule compliance.
- The relationship of the proposed costs to the required work. In this assessment, the engineer/technical member will determine whether all costs proposed are necessary for the satisfactory completion of the work. Proposed costs for work determined to be unnecessary should be excluded.
- The degree to which proposed effort if duplicated. A given cost proposal may contain costs that have been proposed elsewhere in the same proposal or in prior proposals for work that was completed prior to the instant contract.
- The validity of the estimating techniques employed in the proposal. If historical data is used in projecting future cost, the engineer/technical member should determine if the current contract schedule, work load, and other conditions have been adequately considered as a basis for projecting the historical costs to the future.
- The impact of schedule and work load. This evaluation looks at the time period for the contract scope of work and attempts to determine whether the total quantity of effort proposed is correlated with that scope. The team will review the cost proposal to assure themselves there is a proper balance of manpower working on a task versus the time span over which the task is performed.

ENGINEERING (QUANTITATIVE) ANALYSIS-LABOR

Equally important, the buyer will need the engineer/technical member to provide an analysis of the offeror's quantities of labor, materials, and equipment so that buyer (or auditor)-developed rates can be applied against them in deriving a total cost position for negotiation. This analysis of the supplier's costs proposal should determine the following:

- The appropriateness of the proposed skill level and mix. This analysis is important because labor is generally a significant element of cost in most contracts. Part of understanding and evaluating estimated labor is to recognize the patterns in the incidence of different types of labor. Each phase of a work effort will have its own unique combination of required labor types. The skill, grade, and salary levels proposed must make sense when the phases are compared not only with the job as a whole, but with each other.

- The reasonableness of proposed direct labor hours. This analysis attempts to determine whether the supplier has based the hours on proper planning and that it contemplates the sound use of labor and reasonable economy and efficiency of operation. The tests for labor hour reasonableness will include considering the necessity of the proposed effort, the adequacy of the work plan, whether any work has been duplicated, the applicability of historical data, the conditions under which the work will be performed, the estimating methods employed, and the supplier's knowledge of the task.

USING THE LEARNING CURVE

If the supplier has performed similar work for the buyer's organization or similar organizations, the supplier's personnel may manifest a classical learning curve, meaning their productivity will be greater than average on a current contract because they learned how to do the job on earlier contracts. Learning is a universal phenomenon that applies to most contracts (manufacturing more than others). The learning phenomenon is based on the known fact that individuals performing repetitive tasks exhibit a rate of improvement due to increased dexterity. Mental and muscular adjustments made by an individual from the time he or she performs a task for the first time to the time he or she has repeated it a number of times result in a reduction in the time required for each repetition of a uniform unit of work. Not only industrial engineers, but also teachers, psychologists, and others have used this principle for a long time. In the manufacturing arena, "learning" can take place through a combination of factors, including changes in the worker's environment, changes in the flow process, work simplification, engineering changes, changes in work setup, as well as increases in manual dexterity. Because many of these factors are changed as a direct result of management action, many people prefer to think of the learning curve as an "improvement" or "experience" curve. Because learning curve is so well known, we shall continue to use that term. The learning curve can take two different forms. The most common form is called the "unit learning curve." This is nothing more than the curvilinear plot of production volume on the horizontal or X axis against direct labor hours per unit on the vertical or Y axis. Another form, less commonly used, is called the "cumulative average curve." This is nothing more than the curvilinear plot of production volume on the horizontal or X axis against the average cost per unit on the vertical or Y axis. Both of these plots graphically portray and represent the following facts:

- Learning takes place in a constant and predictable manner, namely that when the number of units produced is doubled, the number of man-hours required for the doubled unit is less than the number of man-hours required for the undoubled unit by a percentage called the rate of learning.
- The greater the amount of human involvement in the manufacturing process, the greater the potential rate of learning.
- The learning process occurs only so long as there is constant pressure to reduce man-hours.
- The production of each new design is unique, always starting at the beginning of the learning curve with unit number one.

Curvilinear plots are much more meaningful and helpful for projection purposes when they can be "straightened out" into a straight line. For learning curves, this straightening out process takes place when the equal-scaled curvilinear data is transferred to a logarithmic scale. Learning curves on regular graph paper become straight lines on logarithmic or "log-log" graph paper. With a straight line representation, it is possible to derive rates of learning and to project learning into future periods of production for estimating purposes. On the logarithmic plot for the learning curve, the rate of learning is a derivative from the "slope of the learning curve." The slope of the learning curve is determined graphically and mathematically by picking two points on the unit learning curve, with the first point being x units (take 10 units as an example) and the second point being $2x$ units (20 units in this example). For the tenth unit and the twentieth unit, the labor hours per unit is read from the Y axis. The value for the twentieth unit is placed in the numerator of the equation while the value for the tenth unit is placed in the denominator. The fraction that results is the slope of the learning curve. The rate of learning is then computed by subtracting the slope of the learning curve (always a fraction less than 1.0) from 1.0. Most rates of learning fall between .10 and .30 (10 percent to 30 percent).

The learning curve makes an excellent forecasting or estimating tool for both the supplier and the buyer. By extending the line into future planned production, the supplier and buyer can read from the line the number of labor hours needed for future units to be produced.

The case study entitled "Learning Company" presents a unit learning curve situation as well as the solutions to the questions posed. Readers who want to test their knowledge of unit learning curves should attempt the solutions before reading the answers on the curvilinear and log-log plots. The "Experience Company" case study illustrates the learning or experience phenomenon by using a mathematical approach centered around use of tabular learning curve data to compute cost data for different units in the production process. In the "Experience Company" case study, both unit and cumulative average computations will be shown.

Learning Company

Jim Learner, buyer with People's University, has received a purchase request for some heavy duty air conditioners for some of his older dormitories and classrooms. He will probably need to go sole source to Freeze's, an air conditioner manufacturer that can provide the specialized size and operating needs of the University. Jim knows that Freeze's has an automated manufacturing facility, and that the company's manufacturing costs have been going down over the years. Because one of its university trustees sits on the Board of Directors for Freeze's, Jim was able to obtain the unit production data shown in Table 6-1.

Jim believes Freeze's production operation reflects a classical learning curve. He is interested in plotting the data and extending that data into the production range where his air conditioners will be manufactured. (His air conditioners will be production units 161 to 200).

Discussion

Jim first attempts the plot on a regular geometric scale. The result is shown in Figure 6-1.

The curvilinear plot in Figure 6-1 is interesting in that it presents the classical learning phenomenon in a form that is easy to understand. Notwithstanding this quality, it does not present a form that readily lends itself to projections into future unit production. In extending the line from the 160th to the 200th unit, one can easily draw a line that curves too much (or too little). The result would be a gross misestimate of the man-hours required for those units of production beyond the 160th unit.

Jim needs a plot of the same data on logarithmic paper. Because logarithmic paper is scaled in units of ten rather than units of one, the data plot on

TABLE 6-1

Units Produced	Man-hours per Unit
5	10,000
10	7,500
20	5,625
40	4,219
80	3,164
160	2,373

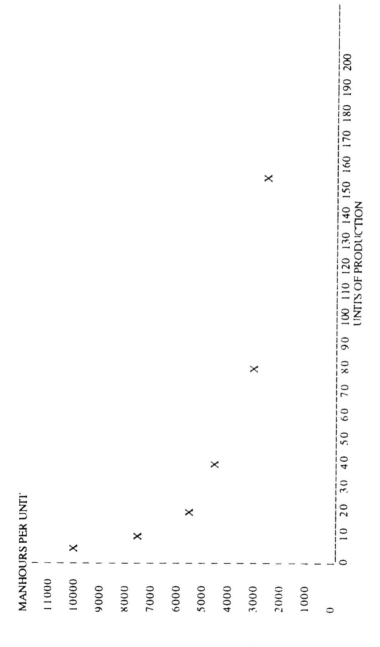

FIGURE 6-1. Plot of learning curve on a regular geometric scale.

this type of paper results in a straight line. A plot of the same data results in the log-log paper plot shown in Figure 6-2.

Jim can use his log-log straight line learning curve and extend that line to the right into the area of future production. By doing this, Jim can read the value of 220 (man-hours), which is the value of the Y or man-hours axis for the 200th unit on the X axis. Jim can also compute this value once he establishes the slope of the learning curve. The slope of a log-log learning curve is always the value of a "doubled" unit divided by the value of its "undoubled" unit. In our example, this would be 7,500 (the man-hour value for the 10th unit) divided by 10,000 (the man-hour value for the 5th unit). The result would, of course, be .75. One could just as easily have used 4,219 (the man-hour value for the 40th unit) and divide it by 5,625 (the man-hour value for the 20th unit). The result would be the same, or .75. By knowing the slope, Jim could determine the value for the 100th unit and then multiply that by .75 to determine the value for the 200th unit. Reading the value from the line would result in a value of 2,900 (man-hours) for the 100th unit. Multiplying this by .75 would yield a value of (2,900) (.75) = 2175 (man-hours). The result would, of course, be much more exact if Jim knew the man-hour history for the 100th unit from the actual production records. A very simple solution to the problem at hand (determining the man-hours for the 161st through 200th units) would be to take the value for the 160th unit and the estimated value for the 200th unit and average them, and then multiply the result by 40 (units). This would yield the following: (2,373 + 2,175)/2 = 2,275 × 40 = 9,100 man-hours. This would be an appropriate value for entering into negotiation with Freeze's for the desired heavy duty air conditioners.

Experience Company

Jim Learner is tired of using learning curves. He wants to accomplish the same objective using learning curve factor tables. He decided to develop some tables based on his knowledge of logarithms. He wants to try out the tables by computing some practice numbers using typical problems he has encountered in past learning curve applications.

In the first situation, Jim wants to use the tabular learning curve data in Appendix A to determine the cost for the 40th unit of production, using a unit 1 cost of $12,000. He wants to find the result assuming several different learning curve slopes. (Jim was provided some additional information from Freeze's, which caused him to reestimate the slope to fall between .79 and .84).

Discussion
Jim knows that he can find the cost of the 40th unit by multiplying a unit 1 cost of $12,000 times the unit factor of the desired unit. In order to assist in that process, he develops the data in Table 6-2.

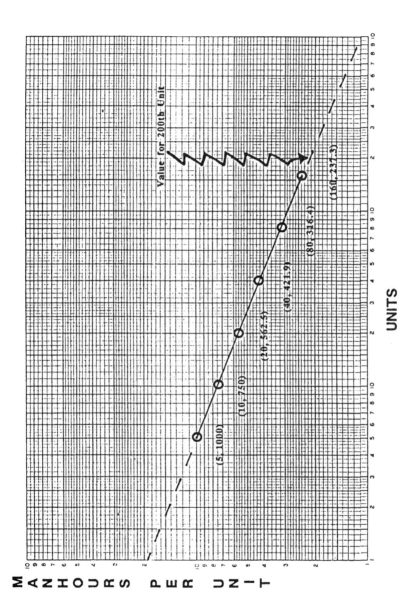

FIGURE 6-2. Plot of learning curve data on log-log paper.

Performing Cost Analysis   103

TABLE 6-2

1. Experience curve	84.0	83.0	82.0	81.0	80.0	79.0
2. Unit factor for unit 40	———	———	———	———	———	———
3. Cost of unit 1 X	$12,000	$12,000	$12,000	$12,000	$12,000	$12,000
4. Cost of unit 40						

From the Appendix A tables, under 84.0, Jim reads down the Unit column until he sees the N value of 40 (extreme left-hand column in the table). The N value is 40 because he is looking for the value of the 40th unit. He extracts the value of 0.39538 (rounded to the nearest five digits). He does the same for the other slopes (83.0, and so forth) and enters them as shown in Table 6-3.

The next step is for Jim to multiply the factors in line 2 times the costs in line 3 and enter the results as shown in Table 6-4.

In the second situation, Jim wants to use the tables in Appendix A to determine the cumulative cost per unit of production, using a unit 1 cost of $4,000. He wants to find the result assuming several different learning curve slopes.

Discussion

Jim knows that he can find the cumulative cost per unit by multiplying a unit 1 cost of $4,000 times the cumulative average factor of the desired number of units. In order to assist in that process, he develops the data in Table 6-5.

From the Appendix A tables, under 84.0, Jim reads down the Cum. Avg. column until he sees the N value of 40 (extreme left-hand column in the

TABLE 6-3

1. Experience curve	84.0	83.0	82.0	81.0	80.0	79.0
2. Unit factor for unit 40	0.39538	0.37097	0.34780	0.32581	0.30497	0.28522
3. Cost of unit 1 X	$12,000	$12,000	$12,000	$12,000	$12,000	$12,000
4. Cost of unit 40	———	———	———	———	———	———

TABLE 6-4

1. Experience curve	84.0	83.0	82.0	81.0	80.0	79.0
2. Unit factor for unit 40	0.39538	0.37097	0.34780	0.32581	0.30497	0.28522
3. Cost of unit 1 X	$12,000	$12,000	$12,000	$12,000	$12,000	$12,000
4. Cost of unit 40	$4,744.56	$4,451.64	$4,173.60	$3,909.72	$3,659.64	$3,422.64

TABLE 6-5

1. Experience curve	84.0	83.0	82.0	81.0	80.0	79.0
2. Cum. avg. factor— unit 40						
3. Cost of unit 1 X	$4,000	$4,000	$4,000	$4,000	$4,000	$4,000
4. Cum. avg. cost for first 40 units						

table). The N value is 40 because he is looking for the value of the 40th unit. He extracts the value of 0.51280 (rounded to the nearest five digits). He does the same for the other slopes (83.0, and so forth) and enters them as shown in Table 6-6.

The next step is for Jim to multiply the factors in line 2 times the costs in line 3 and enter the results shown in Table 6-7.

In the third situation, Jim wants to use the tables in Appendix A to determine the total cost for a given number of units of production, using a unit 1 cost of $1,000. He wants to find the result assuming several different learning curve slopes.

Discussion

Jim knows that he can find the total cost for a given number of units of production by multiplying a unit 1 cost of $1,000 times the cumulative total

TABLE 6-6

1. Experience curve	84.0	83.0	82.0	81.0	80.0	79.0
2. Cum. avg. factor— unit 40	0.51280	0.49084	0.46970	0.44938	0.42984	0.41106
3. Cost of unit 1 X	$4,000	$4,000	$4,000	$4,000	$4,000	$4,000
4. Cum. avg. cost for first 40 units						

TABLE 6-7

1. Experience curve	84.0	83.0	82.0	81.0	80.0	79.0
2. Cum. avg. factor— unit 40	0.51280	0.49084	0.46970	0.44938	0.42984	0.41106
3. Cost of unit 1 X	$4,000	$4,000	$4,000	$4,000	$4,000	$4,000
4. Cum. avg. cost for first 40 units	$2,051.20	$1,963.66	$1,878.80	$1,797.52	$1,717.52	$1,644.24

Performing Cost Analysis   105

TABLE 6-8

1. Experience curve	84.0	83.0	82.0	81.0	80.0	79.0
2. Cum. total factor— unit 40	——	——	——	——	——	——
3. Cost of unit 1 X	$1,000	$1,000	$1,000	$1,000	$1,000	$1,000
4. Total cost for first 40 units	——	——	——	——	——	——

factor for the desired number of units. In order to assist in that process, he develops the table shown as Table 6-8.

From the Appendix A tables, under 84.0 Jim reads down the Cum. Total column until he sees the N value of 40 (extreme left-hand column in the table). The N value is 40 because he is looking for the value of the 40th unit. He extracts the value of 20.512 (rounded to the nearest three digits). He does the same for the other slopes (83.0, and so forth) and enters them as shown in Table 6-9.

The next step is for Jim to multiply the factors in line 2 times the costs in line 3 and enter the results as shown in Table 6-10.

In the fourth situation, Jim wants to use the tables in Appendix A to determine the total cost for a given lot of units between two specific cumulative unit quantities, using a unit 1 cost of $1,000. He wants to determine the total cost of a lot of 20 units starting at unit 21 and ending with unit 40. He wants to find the result assuming several different learning curve slopes.

TABLE 6-9

1. Experience curve	84.0	83.0	82.0	81.0	80.0	79.0
2. Cum. total factor— unit 40	20.512	19.633	18.788	17.975	17.193	16.442
3. Cost of unit 1 X	$1,000	$1,000	$1,000	$1,000	$1,000	$1,000
4. Total cost for first 40 units	——	——	——	——	——	——

TABLE 6-10

1. Experience curve	84.0	83.0	82.0	81.0	80.0	79.0
2. Cum. total factor— unit#40	20,512	19,633	18,788	17,975	17,193	16,442
3. Cost of unit 1 X	$1,000	$1,000	$1,000	$1,000	$1,000	$1,000
4. Total cost for first 40 units	$20,412	$19,633	$18,788	$17,975	$17,193	$16,442

Discussion

Jim knows that he can find the total cost for a lot of units between two specific cumulative unit quantities by multiplying a unit 1 cost of $1,000 times the difference in cumulative total factors encompassing the desired lot of units. In order to assist in that process, he develops the table shown as Table 6-11.

From the Appendix A tables, under 84.0, Jim reads down the Cum. Total column until he sees the N value of 40 and the N value of 20 (extreme left-hand column in the table). He extracts the values of 20.512 and 11.997, respectively (rounded to the nearest three digits). He does the same for the other slopes (83.0, and so forth) and enters them into the table. He then subtracts the unit 20 factor from the unit 40 factor to obtain the difference. The result is shown in Table 6-12.

The next step is for Jim to multiply the factors in line 4 times the costs in line 5 and enter the results as shown in Table 6-13.

In the fifth situation, Jim wants to use the tables in Appendix A to determine the average cost per unit for a lot of units between two specific

TABLE 6-11

1. Experience curve	84.0	83.0	82.0	81.0	80.0	79.0
2. Cum. total factor—unit 40						
3. Cum. total factor—unit 20						
4. Difference between 2 & 3						
5. Cost of unit 1 X	$1,000	$1,000	$1,000	$1,000	$1,000	$1,000
6. Total cost of lot units 21–40						

TABLE 6-12

1. Experience curve	84.0	83.0	82.0	81.0	80.0	79.0
2. Cum. total factor—unit 40	20.512	19.633	18.788	17.975	17.193	16.442
3. Cum. total factor—unit 20	11.997	11.602	11.219	10.847	10.485	10.134
4. Difference between 2 & 3	8.515	8.031	7.569	7.128	6.708	6.308
5. Cost of unit 1 X	$1,000	$1,000	$1,000	$1,000	$1,000	$1,000
6. Total cost of lot units 21–40						

Performing Cost Analysis   107

TABLE 6-13

1. Experience curve	84.0	83.0	82.0	81.0	80.0	79.0
2. Cum. total factor—unit 40	20.512	19.633	18.788	17.975	17.193	16.442
3. Cum. total factor—unit 20	11.997	11.602	11.219	10.847	10.485	10.134
4. Difference between 2 & 3	8.515	8.031	7.569	7.128	6.708	6.308
5. Cost of unit 1 X	$1,000	$1,000	$1,000	$1,000	$1,000	$1,000
6. Total cost of lot units 21–40	$8,515	$8,031	$7,569	$7,128	$6,708	$6,308

cumulative unit quantities, using a unit 1 cost of $1,000. He wants to find the result assuming several different learning curve slopes.

Discussion

Jim knows that he can find the average cost per unit for a lot of units between two specific cumulative unit quantities by dividing the total cost of the lot (as determined in the previous situation) by the total number of units in the lot. In order to assist in that process, he develops the data shown as Table 6-14.

The next step is for Jim to divide the total cost in line 2 by the number of units in line 3 and enter the results as shown in Table 6-15.

In the sixth situation, Jim wants to use the tables in Appendix A to determine the cost of unit 1, given the cost of a specific unit. Jim knows the

TABLE 6-14

1. Experience curve	84.0	83.0	82.0	81.0	80.0	79.0
2. Total cost of lot units 21–40	$8,515	$8,031	$7,569	$7,128	$6,708	$6,308
3. Number of units in lot	20	20	20	20	20	20
4. Average cost per unit of the lot	———	———	———	———	———	———

TABLE 6-15

1. Experience curve	84.0	83.0	82.0	81.0	80.0	79.0
2. Total cost of lot units 21–40	$8,515	$8,031	$7,569	$7,128	$6,708	$6,308
3. Number of units in lot	20	20	20	20	20	20
4. Average cost per unit of the lot	$425.75	$8,401.55	$378.45	$356.40	$335.40	$315.40

108   Cost/Price Analysis: Tools to Improve Profit Margins

cost for unit 25 is $555. He wants to find the result assuming several different learning curve slopes.

Discussion
Jim knows that he can find the cost of unit 1 by dividing the cost of the known unit (25) by the unit factor of the known unit. In order to assist in that process, he develops the table shown as Table 6-16.

From the Appendix A tables, under 84.0, Jim reads down the Unit column until he sees the N value of 25 (extreme left-hand column in the table). He extracts the value of 0.44500 (rounded to the nearest five digits). He does the same for the other slopes (83.0, and so forth) and enters them into the table. He then divides the cost of unit 25 by the respective unit factors. The result is shown in Table 6-17.

In the seventh situation, Jim wants to use the Appendix A tables to determine the cost of unit 1, given the cumulative average cost through a specific number of units. Jim knows the cumulative average cost through unit 25 is $444. He wants to find the result assuming several different learning curve slopes.

Discussion
Jim knows that he can find the cost of unit 1 by dividing the cumulative average cost of unit 25 by the cumulative average factor for unit 25. In order to assist in that process, he develops the data shown as Table 6-18.

From the Appendix A tables, under 84.0, Jim reads down the Cum. Avg. column until he sees the N value of 25 (extreme left-hand column in the table). He extracts the value of 0.57081 (rounded to the nearest five digits). He does the same for the other slopes (83.0, and so forth) and enters them

TABLE 6-16

1. Experience curve	84.0	83.0	82.0	81.0	80.0	79.0
2. Cost of unit 25	$555	$555	$555	$555	$555	$555
3. Unit factor for unit 25	—	—	—	—	—	—
4. Cost of unit 1	—	—	—	—	—	—

TABLE 6-17

1. Experience curve	84.0	83.0	82.0	81.0	80.0	79.0
2. Cost of unit 25	$555	$555	$555	$555	$555	$555
3. Unit factor for unit 25	0.44500	0.42093	0.39789	0.37585	0.35478	0.33465
4. Cost of unit 1	$1,247	$1,318	$1,395	$1,477	$1,564	$1,658

Performing Cost Analysis 109

TABLE 6-18

1. Experience curve	84.0	83.0	82.0	81.0	80.0	79.0
2. Cum. Avg. cost at unit 25	$444	$444	$444	$444	$444	$444
3. Cum. avg. factor for unit 25	——	——	——	——	——	——
4. Cost of unit 1	——	——	——	——	——	——

into the table. He then divides the cumulative average cost at unit 25 by the respective cumulative average factors. The result is shown in Table 6-19.

In the eighth situation, Jim wants to use the Appendix A tables to determine the cost of unit 1, given the cumulative total cost through a specific number of units. Jim knows the cumulative total cost through unit 25 is $22,220. He wants to find the result assuming several different learning curve slopes.

Discussion

Jim knows that he can find the cost of unit 1 by dividing the cumulative average cost of unit 25 by the cumulative total factor for unit 25. In order to assist in that process, he develops the data shown as Table 6-20.

From the Appendix A tables, under 84.0, Jim reads down the Cum. Total column until he sees the N value of 25 (extreme left-hand column in the table). He extracts the value of 14.270 (rounded to the nearest three digits). He does the same for the other slopes (83.0, and so forth) and enters them

TABLE 6-19

1. Experience curve	84.0	83.0	82.0	81.0	80.0	79.0
2. Cum. Avg. cost at unit 25	$444	$444	$444	$444	$444	$444
3. Cum. avg. factor for unit 25	0.57081	0.55024	0.53032	0.51103	0.49234	0.47426
4. Cost of unit 1	$778	$807	$837	$869	$902	$936

TABLE 6-20

1. Experience curve	84.0	83.0	82.0	81.0	80.0	79.0
2. Cum. total cost at unit 25	$22.220	$22.220	$22.220	$22.220	$22.220	$22.220
3. Cum. total factor for unit 25	——	——	——	——	——	——
4. Cost of unit 1	——	——	——	——	——	——

into the table. He then divides the cumulative total cost at unit 25 by the respective cumulative total factors. The result is shown in Table 6-21.

ENGINEERING (QUANTITATIVE) ANALYSIS-MATERIAL

This analysis addresses the reasonableness of the proposed material types and quantities. Material can be quite significant as a cost element. Ideally, the supplier will have developed consolidated lists or bills of materials based on some takeoff from existing plans and drawings. In that eventuality, the engineer/technical member can readily compare the quantity of proposed material with the quantity of material estimated on the detailed in-house estimate. In the absence of such plans, the supplier may have to rely on historical experience. The supplier's estimate of material quantities should be analyzed by not only comparing against prior contract usage but also by projecting reasonable learning in the materials usage. The learning curve phenomenom applies in general to use of materials as well as labor.

ENGINEERING (QUANTITATIVE) ANALYSIS: OTHER DIRECT COSTS

This addresses the reasonableness of the proposed other direct costs (quantities). Other direct costs are those costs that are specifically identified with a project but do not fall within the classification of direct labor or direct material. Examples include equipment, subcontracts, travel, automatic data processing, consultants, and meetings and conferences. These direct costs are reviewed to determine whether they are properly classified in accordance with the supplier's accounting system, and the backup data in support of the costs are valid, current, and applicable to the work required.

TABLE 6-21

1. Experience curve	84.0	83.0	82.0	81.0	80.0	79.0
2. Cum. total cost at unit 25	$22.220	$22.220	$22.220	$22.220	$22.220	$22.220
3. Cum. total factor for unit 25	14.270	13.756	13.258	12.776	12.309	11.856
4. Cost of unit 1	$1,557	$1,615	$1,676	$1,739	$1,805	$1,874

ENGINEERING (QUANTITATIVE) ANALYSIS: PROFIT OR FEE

Although the engineer/technical member should not be asked to develop a recommended profit or fee, he or she can render an opinion on the inherent technical, management, and cost risk perceived in the work as well as an opinion on the degree to which the supplier is willing to assume that risk. Generally, higher cost estimates (padding of cost, either in quantity or rates) evidence an unwillingness by the supplier to assume risk.

ACCOUNTING ANALYSIS

As previously indicated, the buyer will need to develop rates (labor, material, equipment, overhead, G&A and profit) that can be applied against the quantitative positions taken by the engineer/technical member in deriving a total cost position for negotiation. This rate analysis is the accounting analysis of the supplier's cost proposal. Where the supplier has performed similar work in the past, this rate analysis concentrates on historical rates in evidence in the supplier's books and records.

ACCOUNTING ANALYSIS: LABOR RATES

On contracts with incumbent firms, it is a fairly simple proposition to review historical payrolls and trace specific employees to the proposal. Historical rates will, of course, be extrapolated to reflect salary increases in effect during the period of performance. These increases must, of course, be reviewed in the light of past history and economic reasonableness before they are accepted. On contracts that require the supplier to acquire new employees, the offer letters may be reviewed to determine rates. Lacking this evidence, wage and salary survey information available from the American Management Association, the U.S. Department of Labor, and others can be consulted for reasonable, market-based rates in the area of contract performance. The buyer should generally include the following in his or her analysis:

1. Determine the adequacy and reasonableness of the labor cost data the supplier presents, including the degree to which the supplier is quoting salaries and rates for specific personnel.
2. If the supplier is quoting specifically identified personnel, verify the proposed rates against payroll.
3. If the salary or rates quoted for specifically identified personnel are higher than the existing payroll rates, determine what percentage factors were

used to increase the rates during the period of contract performance. Assess the reasonableness of the percentage increase(s).
4. If the supplier is quoting other than specifically identified personnel, verify the categories of labor quoted against the contract requirements. Assure that the position descriptions and pay scales match the job requirements. Determine that the labor rates quoted for these new personnel are not out of line with wage and salary studies appropriate to the area and industry concerned or with the established policy of the supplier.
5. If labor categories (fabrication, assembly, inspection, etc.) are used in lieu of specifically identified personnel, determine the reasonableness of the method of computing weighted average wage or salary rates for the categories.
6. Determine whether subcontractor or consultant labor is included in the supplier's overhead allocation base. If they are, application of an overhead is appropriate, otherwise not.
7. Assure that the supplier's cost accounting and estimating systems are consistent with respect to the manner with which they treat labor costs. This is important in order to determine that the supplier has not "double dipped" by including labor of a similar type as both direct and indirect costs.
8. If the contract is other than a firm-fixed-price type, assure that the supplier's accounting system can accurately record and account for labor by contract or project.

The "Weighted Average Company" case study illustrates the process of arriving at weighted average wage rates. The "Royal Company" case study illustrates use of weighted average wage rates to arrive at a fair and reasonable labor rate position for negotiation.

Weighted Average Company

Mr. Bob Bear wants to determine the weighted average wage rate for You-Call-We-Haul, one of the service suppliers offering on his transportation services contract. He received a lump sum offer for labor computed by the supplier by taking 400 hours times an average rate of $40 per hour. Mr. Bear isn't happy with that fact. He thinks he is being gouged on the total price. Mr. Bear knows the total staffing by labor category for his potential supplier and the average hourly wage rate for each of the potential supplier's labor types. He also knows the supplier estimated 400 hours of labor will be used on his contract. Mr. Bear believes the mix of labor to be used on his contract approximates the mix of labor available to the supplier. He wants to use this information to develop a weighted average wage rate.

Discussion

Mr. Bear starts by developing a matrix of labor quantities and rates as shown in Table 6-22.

Royal Company

In late 199X, Ms. Lily Manor, buyer for Royal Company, received a purchase request for 120 Extraterrestrial Doppler Systems, a hardware item that forms an integral part of their Extraterrestrial Systems. Delivery was specified at ten per month beginning in May 199X. Production was expected to begin on March 1, 199X and continue for 12 months. Because this was an expedited purchase, Ms. Manor had no other recourse than to use the known source, the Space Company, located in Shanghai, People's Republic of China. Ms. Manor asked Mr. Wu, the marketing manager of Space, to provide her with a proposal for cost analysis.

Mr. Wu submitted the following unit price proposal in Yuon (Y):

Direct material	Y 1,560,000
Direct labor (4,000 hours @ Yuon 798)	Y 3,192,000
Subtotal	Y 4,752,000
Manufacturing overhead (95%)	Y 4,514,400
Total manufacturing cost	Y 9,266,400
General and administrative expense (25%)	Y 2,316,600
Total cost	Y11,593,000
Profit (15%)	Y 1,737,450
Total price	Y13,330,450
Total price for 120 units = 120 × Y13,330,450 = Y1,599,654,000	

Ms. Manor assigned Mr. Steve Cadot, cost analyst, to evaluate the proposal. The industrial engineer, Ms. Emma Right, reported to Mr. Cadot that

TABLE 6-22

Labor category	# of Laborers	Avg. Wage/Hr.	Weighted Wage Rate
Driver	100	$ 50.00	$ 5,000
Assistant driver	200	$ 20.00	$ 4,000
Dock worker	400	$ 25.00	$10,000
Materielman	300	$ 30.00	$ 9,000
Total force	1,000		
Total wages		$125.00	
Total force times wages			$18,000
Weighted average wage rate = $18,000/1,000 = $18.00			
Unweighted average wage rate = $125.00/4 = $31.25			

114   Cost/Price Analysis: Tools to Improve Profit Margins

the direct material quantities and direct labor hours were "just about right" and that the manufacturing technique, although it dated back to the WWII era, was adequate for the purpose. The auditor for Space Company (the firm of Calder and Williams, CPAs and management consultants) reported that the overhead and G&A rates for the period in question were estimated at the same rates proposed in the cost proposal. Mr. Cadot thought the direct labor rates were probably above the average on other purchases and probably above comparable rates in China.

Mr. Cadot asked Mr. Wu to provide the following three items in regard to direct labor over the past three years:

1. departmental wage rates and the percentage of that department to the total plant;
2. proposed wage increases; and
3. the employment level within the plant.

Mr. Wu submitted the information shown in Tables 6-23 and 6-24.

Mr. Cadot would like to use this information to verify the plantwide wage rates for 199X; to determine whether the proposed Y798 per hour wage rate is reasonable; and lastly, what a reasonable negotiation position would be for Royal Company.

Discussion

Mr. Cadot can use the Table 6-23 data provided by Mr. Wu to compute a weighted average wage rate as of February 199X as shown in Table 6-25.

Weighted average wage rate = Y4,426,640/6,200 = Y713.97. This confirms the proposed plantwide wage rate for February 199X.

To answer the second question, Mr. Cadot can escalate the Y714 weighted average rate by Y7 per month through February of 199X + 1 (the completion

TABLE 6-23   Departmental wage rates and percentage of total plant employment, February 199X.

1 Department	2 Wage Rates	3 %
Fabrication	Y1,100	5.73%
Assembly	Y 700	44.35%
Foundry	Y 600	19.67%
Machine shops	Y 740	11.45%
Setup	Y 800	11.45%
Take-down	Y 628	7.34%
	Total	100%

Performing Cost Analysis   115

TABLE 6-24   Average monthly direct labor hour rates and number of direct workers.

Month	Year 199X-1			Year 199X	
	Rate	Avg. Hourly # of Workers		Rate	Avg. Hourly # of Workers
January	Y643	10,200		Y713	6,200
February	Y647	9,900		Y714	6,200
March	Y652	9,700		———	———
April	Y660	9,300		———	———
May	Y671	8,100		———	———
June	Y673	7,800		———	———
July	Y677	7,600		———	———
August	Y680	7,400		———	———
September	Y686	7,000		———	———
October	Y692	6,600		———	———
November	Y698	6,400		———	———
December	Y699	6,300		———	———

TABLE 6-25

1	2	3	4	5
Department	Wage Rates	%	# of Workers	Weighted Wages (2X4)
Fabrication	Y1100	5.73%	355	Y 390,500
Assembly	Y 700	44.35%	2,750	Y1,925,000
Foundry	Y 600	19.67%	1,220	Y 732,000
Machine shops	Y 740	11.45%	710	Y 525,400
Set-up	Y 800	11.45%	710	Y 568,000
Take-down	Y 628	7.34%	455	Y 285,740
	Total	100%	6,200	Y4,426,640

of the production period). He can then compute an average for the 12 months ending February 199X to develop a recommended rate for the contract as shown in Table 6-26.

The twelve month average for the period ending February 199X + 1 is Y(721 + 728 + 735 + 742 + 749 + 756 + 763 + 770 + 777 + 784 + 791 + 798)/12 = Y9,114/12 = Y760. Ms. Manor and Mr. Cadot would be inclined to use Y760 as an acceptable hourly rate in lieu of the Y798 proposed.

Ms. Manor and Mr. Cadot can develop a reasonable proposal by substituting a wage rate of Y760 per hour instead of Y798 into the cost proposal as follows:

Direct material	Y 1,560,000
Direct labor (4,000 hours @ Yen 760)	Y 3,040,000
Subtotal	Y 4,600,000
Manufacturing overhead (95%)	Y 4,370,000
Total manufacturing cost	Y 8,970,000
General and administrative expense (25%)	Y 2,242,500
Total cost	Y11,212,500
Profit (15%)	Y 1,681,875
Total price	Y12,894,375

Total price for 120 units = 120 × Y12,894,375 = Y1,545,325,000

ACCOUNTING ANALYSIS: MATERIAL PRICES

The issue here for the buyer is the reasonableness of proposed material prices. On contracts for follow-on work, it is a fairly simple proposition to review books and records to track prices paid for proposed materials. These prices paid, if used as a basis for the estimate, would need to be extrapolated to the period of contract performance using an appropriate wholesale price index escalator. On contracts with no previous incumbent experience, proposed material prices should be pegged to whatever published, catalog, or market prices available in the literature. In conducting the material price analysis, the buyer or auditor must assure that costs are consistently treated

TABLE 6-26

Month	Year 199X Avg. Hourly Rate	Year 199X + 1 Avg. Hourly Rate
January	Y713	Y784 + 7 = Y791
February	Y714	Y791 + 7 = Y798
March	Y714 + 7 = Y721	
April	Y721 + 7 = Y728	
May	Y728 + 7 = Y735	
June	Y735 + 7 = Y742	
July	Y742 + 7 = Y749	
August	Y749 + 7 = Y756	
September	Y756 + 7 = Y763	
October	Y763 + 7 = Y770	
November	Y770 + 7 = Y777	
December	Y777 + 7 = Y784	

Performing Cost Analysis 117

in accordance with the normal cost-keeping system of the supplier, that costs are traceable to and can be supported by such documentation as bills of material, vendors' quotes, and subcontracts; and that costs are reasonable in view of actual prices, with appropriate adjustment for trade discounts, refunds, rebates, allowances, prompt payment, and so forth.

The buyer should generally include the following:

1. Determine the adequacy and reasonableness of the material price data the supplier presents. This should, in most cases, require the supplier to submit a detailed, consolidated price bill of material (BOM).
2. Depending on the number of line items in the BOM, either review the entire BOM or take a random or stratified sample of items from the BOM.
3. If the supplier is quoting material used on previous jobs and projects, verify the proposed prices against vendor invoices for material purchased.
4. If material prices quoted are higher than existing prices available in supplier records, determine what percentage factor(s) was/were used to increase the prices during the period of contract performance. Assess the reasonableness of the percentage increase(s).
5. If the supplier is quoting material not previously used on jobs or projects, verify the proposed prices against contracts, purchase orders, quotations, or catalog information available from the supplier's records or from catalog or market data obtained from independent sources.
6. Determine the degree to which the supplier is obtaining competition on purchase orders for materials.
7. Determine the degree to which the supplier has reduced material prices quoted by discounts, rebates, and allowances.
8. Determine the reasonableness of spoilage and scrap factors.
9. Assess the degree to which the supplier's inventory and material management systems (MMS) provide accurate cost accounting and material control measures. Special attention must be given if the supplier uses an automated MMS.
10. Assure that the supplier's cost accounting and estimating systems are consistent with respect to the manner with which they treat material costs. This is important in order to determine that the supplier has not double dipped by including material of a similar type as both direct and indirect costs.
11. If the contract is other than a firm-fixed-price type, assure that the supplier's accounting system can accurately record and account for material by contract or project.

ACCOUNTING ANALYSIS: OTHER DIRECT COSTS

This issues addresses the reasonableness of proposed other direct cost prices or rates. Other direct costs generally include a combination of different types of costs, including specialized labor, equipment, and support-type costs. The rates for these types of costs should be analyzed by pegging them wherever possible to the market as well as past history and experience by the service supplier and/or the buyer's organization. The buyer should generally include the following in his or her analysis:

1. Determine the adequacy and reasonableness of the other direct cost data the supplier presents.
2. If the supplier has proposed using its own equipment, access to its property records and depreciation schedules will suffice to determine if reasonable amounts have been proposed.
3. If the supplier is leasing equipment, established rate schedules, available from Dataquest, the Association of General Contractors, and the U.S. Army Corps of Engineers can be consulted to peg the proposed rates to "market."
4. Require the prime supplier to review and analyze subcontractor and consultant cost proposals and the documentation supporting them. Make the prime supplier prove the fairness and reasonableness of the subcontractor/consultant proposals. Assure the prime supplier uses competitive source selection procedures in selecting his subcontractors/consultants.
5. Determine the reasonableness of travel, lodging, and meal expenses by assessing the degree to which such costs are consistent with the supplier's established travel and/or relocation policy; are based on the best available air coach and car rental rates; reflect mileage rates consistent with IRS and Department of Transportation guidelines; and reflect reasonable subsistence costs in line with guidelines contained in the Federal Travel Regulation.
6. Determine the reasonableness of ADP and MIS costs by comparing the proposed rates with established commercial or standard rates.
7. Determine the reasonableness of consulting costs by determining if the types of services provided are consistent with established buying organization policy, by investigating to determine whether the supplier's employees are performing as consultants, by benchmarking the consultant rates against other known consulting rates, and by determining whether indirect costs have been applied to consultants' costs. (Unless the

supplier's accounting system provides otherwise, such costs generally merit only a portion, if any, of the normal overhead rates.)
8. Determine the reasonableness of meeting and conference costs by investigating the direct relationship between attendance at the meetings and contract work requirements and by comparing such costs against a commercial or market standard.
9. Assure that the supplier's cost accounting and estimating systems are consistent with respect to the manner they treat these types of cost. This is important to determine that the supplier has not double dipped by including costs of a similar type as both direct and indirect costs.
10. If the contract is other than a firm-fixed-price type, assure that the supplier's accounting system can accurately record and account for these types of cost by contract or project.

ACCOUNTING ANALYSIS: OVERHEAD AND G & A RATES

The reasonableness of proposed overhead and general and administrative (G & A) expense rates should be a matter of some concern to the buyer. If the supplier is doing business with a governmental entity, the chances are that the supplies will have been subjected to some sort of overhead rate audit by that government entity. Audit results are formalized into a rate agreement that tells the firm what rates will be used for prospective bidding purposes as well as for retrospective (close-out) purposes. The buyer should ask the supplier for a copy of its latest governmental rate agreement. Failing that, the buyer should request the supplier divulge its detailed estimate of the costs included in the overhead and G & A pool projections for the contract period in question, divulge estimated bases used in calculating rates for that period, and explain how the rates were derived. Failing that, the buyer should request certified financial information from the company (or obtain it from public records, if the supplier is a public company), which can be used to derive rate approximations.

Cases illustrating the computation of material overhead and manufacturing overhead by a "typical manufacturing firm" are illustrated in the case studies entitled "Raddenberry Company" and "Raddenberry Company II." The "Company A/Company B" case study demonstrates the differential effects of using direct labor cost and direct labor hours as bases for overhead allocation. The "Great Lakes Company" case study demonstrates the graphic analysis of a firm's G & A overhead. The "Banner Company" case study demonstrates the computation of depreciation costs using several different methods of depreciation

120   Cost/Price Analysis: Tools to Improve Profit Margins

Raddenberry Company

Jack Wolf is a buyer for the Next Generation Company. He wants to buy some motors from the Raddenberry Company in Hollywood, California. He knows Raddenberry makes good motors but he isn't sure about the $500 price tag. He has heard from other buyers of motors that the price is too high because of the inefficient materials management organization in Raddenberry. Jack feels it is necessary to understand Raddenberry's material overhead to arrive at an opinion on the degree to which the overhead is reasonable. In response to a request from Jack, Cal Packard, marketing manager for Raddenberry, provided the information shown in Table 6-27.

Discussion

A material overhead rate of 6.48 percent may or may not be reasonable. In order to arrive at an informed opinion on this rate, Jack must determine whether the $700K is a reasonable amount, considering the volume of purchases expected to be processed in 199X. Jack will need to know what the "comparables" are for similar companies. In addition, he will need to know what other overhead costs are for the Raddenberry Company. Many companies will include their material overhead expenses in manufacturing overhead or even G & A overhead. In short, Jack will need to take both a broader and a narrower focus on this rate to assess its reasonableness. The data needed for this analysis are not readily available from the case material.

Raddenberry Company II

Assume that Raddenberry has three production departments and two departments that support the production departments. The Raddenberry policy is to allocate supporting department costs to the production departments based on maintenance labor hours expected to be expended in support of the

TABLE 6-27

Material Overhead Cost	199X Estimated Costs
Purchasing department	$ 200K
Receiving department	$ 160K
Material storage	$ 120K
In-plant material handling	$ 140K
Freight-in	$ 80K
Total Estimated Cost	$ 700K
Estimated purchase volume for 199X:	$10,800K
Estimated material overhead rate: $700K/$10,800 = .0648 or 6.48%	

three production departments and the supporting inventory storage costs based on square feet of storage space needed for production materials. After allocating supporting department costs, Raddenberry computes a manufacturing overhead based on allocating total factory overhead expenses to direct labor dollars. This is illustrated in Table 6-28.

Assume the supporting departments will need to allocate 60 percent of maintenance hours to production department A, 25 percent of maintenance hours to production department B, and 15 percent of maintenance hours to production department C. Assume further that the supporting departments will allocate 55 percent of storage space to production department A, 40 percent of storage space to production department B, and 5 percent of storage space to production department C. This results in the cost allocations shown in Table 6-29.

Assume production department A has estimated 4,000 direct labor hours (at a weighted average of $22 per hour) for the period in question, production department B has estimated 3,000 direct labor hours (at a weighted average of $25 per hour), and production department C has estimated 2,000 direct labor hours (at a weighted average of $30 per hour). Overhead rates, computed by dividing total overhead expenses by direct labor dollars, can be computed as shown in Table 6-30.

Company A/Company B

Company A and Company B are two competitors in the widget manufacturing industry. Company A and Company B are virtually identical. They have estimated the same direct labor dollars for the next fiscal year, pay the same weighted average hourly rate per direct labor dollar, and have estimated the same overhead costs for the next fiscal year. The only thing that differentiates them is that they allocate their overhead on different bases. Company A

TABLE 6-28 Estimated overhead costs for Raddenberry Company maintenance inventory storage, and production departments

Expense Category	Company TTL	Maint.	Invent.	Production Departments		
				Dept A	Dept B	Dept C
Indirect labor	$10,000	$1,800	$1,500	$ 3,000	$2,000	$1,700
Depreciation	$ 8,500	$1,500	$1,500	$ 2,000	$1,800	$1,700
Insurance, taxes, rent, heat, power, and light	$16,000	$2,000	$2,000	$ 5,000	$4,000	$3,000
Subtotal	$34,500	$5,300	$5,000	$10,000	$7,800	$6,400

TABLE 6-29 Allocation of estimated supporting department overhead costs to production departments

Expense category	Maint.	Invent.	Production Department		
			Dept. A	Dept. B	Dept. C
Maintenance			$3,180	$1,325	$795
			.60×$5300	.25×$5300	.15×$5300
Inventory			$2,750	$2,000	$250
			.55×$5000	.40×$5000	.05×$5000

TABLE 6-30 Computation of overhead rates

Production department A:	Total Overhead Expenses ($10,000 + $3,180 + $2,750) Direct Labor Cost (4,000 × $22) = $15.930/$88.000 = 18.1%
Production department B:	Total Overhead Expenses ($7,800 + $1,325 + $2,000) Direct Labor Cost (3,000 × $25) = $11,125/$75,000 = 14.8%
Production department C:	Total Overhead Expenses ($6,400 + $795 + $250) Direct Labor Cost (2,000 × $30) = $7,445/$60,000 = 12.4%

allocates based on direct labor hours and Company B allocates based on direct labor dollars. Observe how this affects their cost estimating for a job that requires the same amount of direct labor hours and direct labor rate. Assume the information shown in Table 6-31.

Assume further that both companies get a request for proposals for a job they both estimate at 3,000 hours at a weighted average weight of $40.00 per hour (Table 6-32). What will each company propose for the work?

Discussion

Based on their different methods allocating overhead, Company A would get the job, since the total proposal was $3,600 lower than the proposal from Company B. The moral of the story of this case is that direct labor dollars are not a good base when the labor required for a specific job deviates substantially from the estimated companywide weighted average labor rate ($40 versus $25 per hour).

TABLE 6-31

	Company A	Company B
Estimated direct labor hours for the next fiscal year	500,000	500,000
Estimated weighted average labor rate for the next fiscal year	$25.00/Hr	$25.00/Hr
Estimated overhead expenses for the next fiscal year	$1,000,000	$1,000,000
Basis for overhead allocation	Direct Labor Hours	Direct Labor Dollars
Overhead Rate	$1,000,000/500,000 = $2.00/Hr	$1,000,000/$12,500,000 = .08 = 8%

TABLE 6-32

	Company A	Company B
Estimated direct labor cost for the job	$120,000	$120,000
Estimated overhead expense for the job	$6,000 ($2.00×3,000)	$9,600 (.08×$120,000)
Total estimated cost for the job	$126,000	$129,600

ACCOUNTING ANALYSIS: DEPRECIATION COST

Banner Company

Joe Flag, buyer for the Veterans Administration, wants to buy some flagpoles from the Banner Company. He knows the company will be using four different pieces of recently purchased equipment during the production process and that the production will go on for a five-year period. He has received a cost proposal from the Banner Company, which contains some unbelievable overhead rates. He has heard that the rates are high because Banner has a lot of equipment depreciation built into the rates. In response to a request from Joe, the Banner Company provided the equipment acquisition cost, salvage value, useful life, and applicable depreciation method as shown in Table 6-33.

Joe Flag wants to use this information to compute the annual and total depreciation costs for the four items of equipment.

Discussion

Joe remembers from his college accounting classes that sum-of-the-years digits and declining-balance methods of depreciation are "accelerated"

TABLE 6-33

	Truck	Computer	Drill Press	Robotic Painter
Cost	$30,000	$8,000	$500,000	$1,000,000
Useful life	5 Years	5 Years	5 Years	10,000 Hours
Salvage value	$5,000	0	$50,000	$100,000
Depreciation method	Declining balance	Straight line	Sum-of-years digits	Service-hours
Year:				Hrs. of use
1				1,500
2				1,700
3				2,200
4				2,500
5				2,100
Total				

methods of depreciation, resulting in allocation of depreciation expense early in the life of the item being depreciated. He thinks that is one reason why the depreciation costs are so high. He also remembers that the declining-balance method is sometimes called the double-declining-balance method because twice the straight line rate of depreciation is taken in the early years of the useful life. He also remembers that only the straight line and service-hours methods subtract salvage value before computing the depreciation amounts. Armed with that knowledge, Joe decides the easiest one to compute is the straight line method. He knows the salvage value ($500) is subtracted from the cost ($8,000) to yield a depreciable value of $7,500. He knows also the depreciable value is divided by the number of years of useful life (five) to yield the annual depreciation ($1,500). He thereupon enters this data into the table provided by Banner as shown in Table 6-34.

Joe decides the next easiest one to compute is the service-hour method. He knows the salvage value ($100,000) is not subtracted from the cost ($1,000,000), to yield a depreciable value of $1,000,000. He knows also the depreciable value is divided by the total service hours and then multiplied by the service hour usage to yield the depreciation for each service period. He thereupon enters this data into the table provided by Banner as shown in Table 6-35.

Joe decides the next easiest one to compute is the declining-balance method. He knows the salvage value ($5,000) is not subtracted from the cost ($30,000), to yield a depreciable value of $30,000. He knows also the depreciation rate to be applied against the depreciable value is twice the straight line rate (100/5 or 0.20). Twice this rate is 0.40. After this rate is determined, it is multiplied by the depreciable value (book value minus amount previously

TABLE 6-34

	Truck	Computer	Drill Press	Robotic Painter
Cost	$30,000	$8,000	$500,000	$1,000,000
Useful life	5 Years	5 Years	5 Years	10,000 Hours
Salvage value	$5,000	$500	$50,000	$100,000
Depreciation method	Declining balance	Straight line	Sum-of-years digits	Service-hours
Year:				Hrs. of use
1		$1,500		1,500
2		$1,500		1,700
3		$1,500		2,200
4		$1,500		2,500
5		$1,500		2,100
Total		$7,500		

TABLE 6-35

	Truck	Computer	Drill Press	Robotic Painter
Cost	$30,000	$8,000	$500,000	$1,000,000
Useful Life	5 Years	5 Years	5 Years	10,000 Hours
Salvage value	$5,000	$500	$50,000	$100,000
Depreciation method	Declining balance	Straight line	Sum-of-years digits	Service-hours
Year:				
1		$1,500		1,500
				$150,000
2		$1,500		1,700
				$170,000
3		$1,500		2,200
				$220,000
4		$1,500		2,500
				$250,000
5		$1,500		2,100
				$210,000
Total		$7,500		$1,000,000

depreciated). The first year's depreciation he computes at (.40)($30,000) = $12,000. The second year's depreciation he computes at (.40)($30,000−$12,000) = $7,200. The third year he computes at (.40)($30,000− $12,000−$7,200) = $4,320. Joe looks at the total value of depreciation he has computed through the third year and sees that the total is $12,000 + $7,200 + $4,320 = $23,520. If he continues into the fourth year taking the full amount of depreciation, he would derive the following: (.40)($30,000−$23,520) = $2,592. Ordinarily, he would be inclined to take that full amount, but he remembers from his accounting classes that he cannot take more depreciation than the acquisition cost less the salvage value. With this in mind, he computes the fourth year of depreciation as follows: $30,000 − $5,000 − $12,000 − $7,200 − $4,320 = $1,480. This amount, rather than the $2,592, becomes his fourth-year depreciation. He thereupon enters this data into the table provided by Banner as shown in Table 6-36.

Joe feels he is now ready to tackle the sum-of-the-years digits method. He knows the salvage value ($50,000) is subtracted from the cost ($500,000), to yield a depreciable value of $450,000. He knows also the depreciation rate to be applied the first year against the depreciable value is computed by multiplying a fraction times the depreciable value. The fraction is determined by using as a numerator the number of years remaining in the first year to be depreciated (5) and using as a denominator the sum of the number of years

TABLE 6-36

	Truck	Computer	Drill Press	Robotic Painter
Cost	$30,000	$8,000	$500,000	$1,000,000
Useful Life	5 Years	5 Years	5 Years	10,000 Hours
Salvage value	$5,000	$500	$50,000	$100,000
Depreciation method	Declining balance	Straight line	Sum-of-years digits	Service-hours
Year:				
1	$12,000	$1,500		1,500 $150,000
2	$7,200	$1,500		1,700 $170,000
3	$4,320	$1,500		2,200 $220,000
4	$1,480	$1,500		2,500 $250,000
5	———	$1,500		2,100 $210,000
Total	$25,000	$7,500		$1,000,000

Performing Cost Analysis   127

to be depreciated, determined to be 5 + 4 + 3 + 2 + = 15. After this fraction (⁵⁄15 or .3333) is determined, it is multiplied by the depreciable value (book value minus salvage value). The first year's depreciation he computes at (.3333)($450,000) = $150,000. The depreciation rate to be applied the second year is determined by using as a numerator the number of years remaining after the first year of depreciation (4) and using as a denominator the same sum of digits (15). The second year's depreciation rate then becomes ⁴⁄15 or .2667 and the depreciation becomes (.2667)($450,000) = $120,000. The depreciation rate to be applied the third year he computes at ³⁄15 or .2000. The third year's depreciation then becomes (.2000)($450,000) = $90,000. The fourth-year depreciation rate becomes ²⁄15 or .1333 and the depreciation becomes (.1333)($450,000) = $60,000. The fifth-year depreciation rate, is of course ¹⁄15 or .0667 and the depreciation then becomes (.0667)($450,000) = $30,000. Joe then applies this data to his table as shown in Table 6-37.

SUMMARY

As explained in this chapter, full-blown cost analysis (supported by price analysis) will generally be required for purchasing items or services using complex buyer specifications, for major noncompetitive purchases, for pur-

TABLE 6-37

	Truck	Computer	Drill Press	Robotic Painter
Cost	$30,000	$8,000	$500,000	$1,000,000
Useful life	5 Years	5 Years	5 Years	10,000 Hours
Salvage value	$5,000	$500	$50,000	$100,000
Depreciation method	Declining balance	Straight line	Sum-of-years digits	Service-hours
Year:				
1	$12,000	$1,500	$150,000	1,500 $150,000
2	$ 7,200	$ 1,500	$ 120,000	1,700 $170,000
3	$ 4,320	$ 1,500	$ 90,000	2,200 $220,000
4	$ 1,480	$ 1,500	$ 60,000	2,500 $250,000
5	———	$ 1,500	$ 30,000	2,100 $210,000
Total	$25,000	$7,500	$450,000	$1,000,000

chases for which selection is based on factors other than price, and for use of other than firm-fixed-price or fixed-price-with-economic price adjustment (escalation) type contracts.

In such circumstances the buyer will generally find it necessary to request other functional experts to help conduct engineering and accounting analysis. The dividing line between the two is somewhat easily determined because engineering analysis devotes itself to a quantitative analysis of labor hours, material quantities, and equipment quantities while accounting analysis devotes itself to a rate analysis (labor rates, materials prices, and overhead rates). The engineering analysis should address the completeness of the supplier's proposed costs, the relationship of the proposed costs to the required work, the degree to which proposed effort if duplicated, the validity of the estimating techniques employed in the proposal, and the impact of schedule and work load.

Equally as important, the buyer will need the engineer/technical member to provide an analysis of the offeror's quantities of labor, materials, and equipment so that buyer (or auditor)-developed rates can be applied against them to derive a total cost position for negotiation. This analysis of the supplier's cost proposal should determine the appropriateness of the proposed skill level and mix, the reasonableness of proposed direct labor hours, the reasonableness of the material quantities proposed, and the reasonableness of the equipment quantities (hours or days of use, and so forth).

Once the quantitative analysis is performed of the supplier's cost proposal, accounting analysis is applied to determine the appropriate labor rates, material prices, equipment rates, overhead rates, and G & A rates to be used to monetize the proposed quantities into dollars.

This chapter has addressed several of the analytical techniques used to conduct engineering and accounting analysis, including learning curve analysis, weighted average wage rate analysis, depreciation analysis, and overhead analysis.

7
Assessing Risk and Developing Prenegotiation Profit Objectives

DETERMINANTS OF FAIR AND REASONABLE PROFIT OR FEE

The buyer performing cost analysis must attempt to ensure a fair and reasonable price that includes fair and reasonable costs and a fair and reasonable profit. To determine a fair and reasonable profit or fee, the buyer should assess the technical, management, and cost risk of the work and determine the degree to which the supplier is willing to assume that risk. Generally, higher cost estimates (padding costs, either in quantity or rates) evidence an unwillingness by the supplier to assume risk. In addition to considering technical, management, and cost risk and the relative difficulty of the job, the buyer should consider the size of the job, the period of performance, the amount of investment being made by the supplier in performing the work, the amount of assistance (buyer-provided property and financing) being provided by the buying organization, the amount of subcontracting involved, and very importantly, the type of contract. Most of these factors deal with risk in various manifestations.

METHODOLOGY FOR DEALING WITH PROFIT OR FEE DETERMINANTS

Rather than treat risk in a macro sense, it is better to break down risk into its components and deal with them each as objectively as possible. One method of doing this is to employ a procedure approximating the following:

- Select the factors impacting on the risk of the contract.
- Assign factor weights to each risk factor. The total weight for all risk factors should be 100 percent (the "universe").

- Establish a range of profits for the contract (or alternatively, for each risk factor).

The range of profits for a contract should be established by selecting as the lowest percentage in the range that profit percentage which represents a profit for a very low-risk contract. A cost-plus-fixed-fee contract would be considered very low risk, warranting a low from 0 percent to 4 percent. A firm-fixed-price contract would be considered a higher risk, warranting a low-risk rating from 4 percent to 7 percent. The highest percentage in the range would represent that profit percentage which represents a profit for a very high-risk contract. A cost-plus-fixed-fee contract would probably warrant a high in the 7 percent to 10 percent range. A firm-fixed-price contract would probably warrant a high in the 12 percent to 15 percent range. This discussion is summarized by Table 7-1.

Contracts fitting the risk spectrum between the above two extremes would obviously warrant intermediate profit ranges.

After selecting an appropriate range of profits for the particular type of contract, the buyer should then accomplish the following:

- Apply appropriately selected risk percentages to each of the risk factors.
- Multiply the selected risk percentages by their factor weights.
- Aggregate the resultant weighted profit percentages. The total of all the computations is the recommended profit percentage.

Let's see how we can apply this process to the Plain-O Construction Company case study.

Plain-O Construction Company

In March 199X, the City of Elmora issued a sole-source solicitation for sewer construction services to the Plain-O Construction Company. (The City's construction buyer, Omar Brown, did not know at the time that Plain-O was owned by the Mayor's brother.) The project involved constructing 30 miles of sewer line. Plain-O proposed to subcontract the brush clearing and ditch digging services to small and small-disadvantaged businesses (in order to meet the City's set-aside goals). Plain-O estimated the total job at $28.00M

TABLE 7-1

Type of Contract	Profit Range
Cost-plus-fixed-fee	0% (low) to 10% (high)
Firm-fixed-price	7% (low) to 15% (high)

and wanted to subcontract with The Diggers Company for the ditch digging in the amount of $420,000 and with the Landscaping Company for the brush clearing in the amount of $400,000. Plain-O is required to complete the line in increments, with two miles a month for the first six months of the contract and three miles a month for the last six months. Since the City has very good design specifications (prepared by the Builders Architect/Engineering Company), the City's facilities engineering staff believes a firm-fixed-price contract is appropriate.

When questioned about the capital equipment and financing required for the job, the Plain-O General Manager, Nell Fernandez, explained that the only new facilities and equipment investment contemplated is routine replacement of existing assets and replacing several older dump trucks with new models. Plain-O intends to use its own considerable resources to finance the front end of the job and to bill the City as work progresses. Plain-O is not asking to use any City equipment nor is it asking for advance payments or extraordinary progress payments.

The City's facilities engineering staff has used the cost estimate prepared by Builders Architect/Engineering Company and the considerable talents of the City cost estimator, Sally Evans, to review the Plain-O cost proposal and come up with a city position on cost. This effort resulted in the figures in Table 7-2.

Discussion
Let's work through the above recommended steps in order to arrive at a profit determination.

Select the factors impacting on the risk of the contract. Omar Brown analyzes the contract and determines the following factors impact on the risk of the contract:

1. Technical, management, and cost risk of the job;
2. Relative difficulty of the work;

TABLE 7-2

	Plain-O's Proposal	City's Proposed Cost
Materials	$ 1.55M	$ 1.45M
Subcontracts	$ .82M	$ .75M
Direct labor	$ 7.28M	$ 6.11M
Total direct costs	$ 9.65M	$ 8.31M
General & administrative expense	$13.70M	$11.43M
Total cost	$23.35M	$19.74M
Profit	$ 4.65M (19.9% of cost)	?
Total price	$28.00M	?

132   Cost/Price Analysis: Tools to Improve Profit Margins

3. Size of the job;
4. Period of performance;
5. Supplier's capital investment;
6. Assistance by the buyer; and
7. Amount of subcontracting.

Assign factor weights to each risk factor. Omar Brown assigns the following weights to the above risk factors:

1. Technical, management, and cost risk   20%
2. Relatively difficulty of the work   15%
3. Size of the job   15%
4. Period of performance   15%
5. Supplier's capital investment   5%
6. Assistance by the Buyer   5%
7. Amount of Subcontracting   25%
   Total   100%

Establish a range of profits for the contract (or alternatively, for each risk factor). Omar Brown believes this particular firm-fixed-price contract should be assigned a low risk at .03 and a high risk at .12.

Apply appropriately selected risk percentages to each of the risk factors. Omar Brown recognizes that this is the most important step of all. Although he realizes each risk percentage selection is essentially a subjective assessment, he believes it is possible to objectify the process to a certain degree. He accordingly requests that Sally Evans develop a table of profit factors for various job sizes, periods of performance, and amounts of subcontracting, with .03 being the lowest profit and .12 the highest profit in each case. Ms. Evans provides him with the data found in Tables 1 through 3 at Appendix B.

Brown uses the matrix found in Table 7-3 to organize his data.

Project: Construct 30 miles of sewer line
Contract number: Elmora 0001-0001
Estimated by: Sally Evans
Profit objective for: Plain-O Construction Company

After consulting with his facilities engineering staff and his cost estimator, Omar decides to apply a weight of .075 to technical, management, and cost risk (average risk); a weight of .050 to relatively difficulty of the work (medium-low risk); a weight of .030 to size of the job (from Table 1); a weight of .071 to period of performance (from Table 2); a weight of .050 to supplier

Assessing Risk and Developing Prenegotiation Profit Objectives 133

TABLE 7-3

Factor	Rate	Weight	Value
1. Technical, management, and cost risk	20%		
2. Relatively difficulty of the work	15%		
3. Size of the job	15%		
4. Period of performance	15%		
5. Supplier's capital investment	5%		
6. Assistance by the buyer	5%		
7. Amount of subcontracting	25%		
	Total		

capital investment (medium-low risk); a weight of .120 to assistance by the buyer (none required); and a weight of .120 to amount of subcontracting (from Table 3). Applying these weights results in the figures found in Table 7-4.

Multiply the selected risk percentages by their factor weights. This part of the process results from multiplications of each of the factor rates by its determined weight, as shown in Table 7-5.

TABLE 7-4

Factor	Rate	Weight	Value
1. Technical, management, and cost risk	20%	.075	
2. Relatively difficulty of the work	15%	.050	
3. Size of the job	15%	.030	
4. Period of performance	15%	.071	
5. Supplier's capital investment	5%	.050	
6. Assistance by the buyer	5%	.120	
7. Amount of subcontracting	25%	.120	

TABLE 7-5

Factor	Rate	Weight	Value
1. Technical, management, and cost risk	20%	.075	.015
2. Relatively difficulty of the work	15%	.050	.008
3. Size of the job	15%	.030	.004
4. Period of performance	15%	.071	.011
5. Supplier's capital investment	5%	.050	.002
6. Assistance by the buyer	5%	.120	.006
7. Amount of subcontracting	25%	.120	.030

Aggregate the resultant weighted profit percentages. This part of the process results from adding all the values in the right-hand column as shown in Table 7-6.

Based on this analysis, Omar Brown would apply a profit of 7.6 percent against the City's recommended cost position as shown in Table 7-7.

Based on this, Omar would attempt to negotiate a price of $21.25M with Plain-O.

SUMMARY

The buyer performing cost analysis must attempt to ensure a fair and reasonable price that includes fair and reasonable costs and a fair and reasonable profit. To determine a fair and reasonable profit or fee, the buyer should assess the technical, management, and cost risk of the work and determine the degree to which the supplier is willing to assume that risk. In addition to

TABLE 7-6

Factor	Rate	Weight	Value
1. Technical, management, and cost risk	20%	.075	.015
2. Relatively difficulty of the work	15%	.050	.008
3. Size of the job	15%	.030	.004
4. Period of performance	15%	.071	.011
5. Supplier's capital investment	5%	.050	.002
6. Assistance by the Buyer	5%	.120	.006
7. Amount of Subcontracting	25%	.120	.030
	Total:		.076

TABLE 7-7

	Plain-O's Proposal	City's Proposed Cost
Materials	$ 1.55M	$ 1.45M
Subcontracts	$ .82M	$ .75M
Direct Labor	$ 7.28M	$ 6.11M
Total Direct Costs	$ 9.65M	$ 8.31M
General & Administrative Expense	$13.70M	$11.43M
Total Cost	$23.35M	$19.74M
Profit	$ 4.65M (19.9% of cost)	$ 1.50M (7.6% of cost)
Total Price	$28.00M	$21.24M

considering technical, management, and cost risk and the relative difficulty of the job, the buyer should consider the size of the job, the period of performance, the amount of investment being made by the supplier in performing the work, the amount of assistance (buyer-provided property and financing) being provided by the buying organization, the amount of subcontracting involved, and very importantly, the type of contract. Most of these factors deal with risk in various manifestations. Rather than treat risk in a macro sense, it is better to break down risk into its components and deal with them each as objectively as possible. One method of doing this is to employ a procedure that includes the following:

- Select the factors impacting on the risk of the contract.
- Assign factor weights to each risk factor. The total weight for all risk factors should total 100 percent (the "universe").
- Establish a range of profits for the contract (or alternatively, for each risk factor).
- Apply appropriately selected risk percentages to each of the risk factors.
- Multiply the selected risk percentages by their factor weights.
- Aggregate the resultant weighted profit percentages. The total of all the computations is the recommended profit percentage.

8
Conducting Price Analysis Supplemented by Partial Cost Analysis

SITUATIONS CALLING FOR PRIMARY RELIANCE ON PRICE ANALYSIS AND SECONDARY RELIANCE ON COST ANALYSIS

The buyer will frequently be confronted with situations where price analysis will suffice for determining a fair and reasonable price for most of the value of the buy yet be insufficient for a portion of the total value. This is more common than most would admit. Purchasing situations calling for price analysis supplemented with partial cost analysis include the following:

- Purchasing modified commercial items.
- Purchases from off-shore sources.
- F.O.B. origin purchases.
- Purchases from suppliers unwilling (or unable) to provide complete cost detail.

Purchasing Modified Commercial Items

This situation is very common. Sooner or later, a buyer will be required to purchase an item that is essentially off-the-shelf with some "bells and whistles" to make it unique. Quite often, the basic item will be a catalog or market item, for which pricing will be readily available. The challenge for the buyer is to determine the value of the modifications to the basic commercial item. This value assessment is generally dependent on supplier-submitted cost data for the modifications performed, followed by buyer cost analysis of that limited cost information. Such a pricing approach is not that much different from the approach used to price contract changes and modifications. Both analyses start with a given base item or work element and then add something incremental, which must be priced independently of the base item or work element. When accomplishing this, the buyer must be sure to consider any potential cost reduction in the basic item or work element as a result of the modifications to that base. This may well be the case if the modified commercial item is modified in the basic production process rather than after the commercial item has been produced. In pricing the value of the modifications, the buyer will be required to assess a reasonable profit for the value of the additional work. Normally, the profit rate for the basic item is continued into the cost of the modifications. A more scientific approach would be to separately analyze the risk of the modifications in order to arrive at a profit rate or amount commensurate with that risk.

Purchases From Off-Shore Sources: The "Total Landed Cost"

Buyers from off-shore sources should be very much concerned with their ability to determine the "total landed cost" of the off-shore items. Monczka and Giunipero (1990, p. 43) refer to the process of arriving at the total landed cost as the "Total Cost Sourcing Evaluation Model." The off-shore buyer has a multitude of cost factors to add to the off-shore basic purchase price in order to get the bottom line price. Most quotations in off-shore purchasing are obtained on a cost plus insurance and freight (CIF) basis, which guarantees inclusion of the marine insurance and transportation (to the U.S. port) in the contract price. There are many other costs in an off-shore transaction, some readily apparent and some hidden to the buyer, all of which must either be included in the contract price or added to that price in order to arrive at the total landed cost. The buyer must be adept at evaluating all these different categories of costs not commonly found in a domestic transaction. If something is missed, the buyer may ultimately pay more than a reasonable domestic price for the same or similar items. The problem is complicated by the fact that several elements of cost must be estimated. This raises the possibility that estimates prepared (necessarily) on a precontract basis may later turn into costs substantially higher or lower than the original estimates.

Composition of Total Landed Cost
In addition to the basic item purchase price, the ultimate cost of delivering the goods to the buyer's facility will generally include:

1. Costs of assists, if any are required.
2. Escalation cost, if permitted by the purchase order terms.

3. Cost of special, export packaging appropriate to the item.
4. Transportation from the seller's facility to the port.
5. Port handling costs (wharfage, loading, warehousing, freight forwarding, and so forth) at the port of debarkation.
6. Export processing fees assessed by the exporter's government.
7. Certificate of inspection at the port of debarkation.
8. Ocean shipping costs.
9. Marine insurance premiums.
10. Port handling costs (wharfage, unloading, warehousing, and so forth) at the U.S. port of entry.
11. Customshouse broker fees.
12. Customs duties (depending on the item and country of origin).
13. Inland transportation costs (including demurrage, if appropriate).
14. Financing charges, including bank charges for processing documents and issuance of a letter of credit.
15. Costs of foreign exchange conversion, as appropriate.
16. Hedging (forward or futures contracts, or purchase of currency or commodity futures).

For a CIF shipment, items 1, 2, 3, 4, 5, 6, 8, and 9 will be included in the CIF contract price. Item 7 is normally discretionary on the part of the buyer. The contract for certificate of inspection services is generally between the buyer and the inspection firm, for obvious conflict of interest reasons. Separate buyer contracts are also necessary for the customs broker, for a transportation common or contract carrier within the U.S., for the bank letter of credit, and for the hedging transaction. More contracts necessitate more contracting time by the buyer (and legal fees, if the firm uses outside counsel).

Different INCOTERMS than CIF call for different allocations of cost between the buyer and seller. They also involve a different allocation of responsibilities for handling the different aspects of the international transaction. A summary of the principal INCOTERMS and their respective allocations of responsibilities and costs is shown in chapter 3.

Importers must, in addition, contemplate other costs of buying off-shore. These costs are often difficult to trace or allocate directly to the items being purchased. Most importers account for these costs in their overhead, making it difficult to determine a true total landed cost for foreign purchases. These costs generally are almost always estimated on an incremental basis. Monczka and Giunipero (1990, p. 47) refer to these costs as additional inventory carrying costs and extra communication and documentation costs.

Inventory carrying costs are those costs incurred when holding inventory. Monczka and Giunipero (1990, p. 47) assert that they generally include "interest rate foregone by investing funds; insurance; property taxes; storage;

and obsolescence." One must assume that interest rate foregone means interest/investment income foregone, which can be determined (computed) by multiplying inventory investment times the firm's cost of capital.

Most firms find these costs to be substantial. They generally find that buying off-shore requires a higher inventory level. Additional inventory carrying costs for off-shore purchases can generally be determined by computing estimated carrying costs for domestic items, by computing the equivalent carrying costs for off-shore items, and then subtracting one from the other.

Communication and documentation costs increase due to time and distance from the off-shore supplier. Such costs could include:

1. Telecommunication expenses.
2. Travel, lodging, meals, and miscellaneous expenses for trips to supplier countries.
3. Metrication (English to metric conversion) costs.
4. Increased costs of postage and correspondence.
5. Translation expenses whenever the off-shore supplier insists that the contract be written in the supplier's native language.
6. Increased contract administration and legal time (contracts with off-shore suppliers often require time and effort of higher level personnel not generally involved with domestic transactions).

Typical Example
Let us take a typical off-shore transaction, in this case for a particular printed circuit board built to buyer specification. (See Table 8-1.) Assume the contract is for 10,000 units per month for a period of 12 months from a firm in Taiwan. Terms are CIF, Port of Los Angeles, and the purchasing lead-time from Taiwan is 15 weeks. Assume that the buyer must provide buyer-provided material (assists) amounting to $6,000.

Against this total landed cost, the buyer must compare his or her best domestic quotation. Here too the buyer must make sure the costs for the domestic purchase are fairly and correctly determined. In this age of JIT purchasing the total landed cost for domestic purchases should include little, if any, inventory carrying costs and few, if any, additional administrative costs. The buyer must, however, make sure costs of transportation are added to an F.O.B. Origin quotation in order to make it equivalent to the off-shore total landed cost.

Using Estimates to Make Decisions
Assume the lowest domestic quotation (transportation included) for the circuit board is $5,801,025. To this amount must be added the "assists" of

TABLE 8-1

Total purchase price per month:		
Direct cost of manufacturing	$ 30 × 10,000 units	$300,000
Export packaging	$ 1 × 10,000 units	10,000
Transportation to port of debarkation	$200 per container × 6 containers	1,200
Freight forwarder's fee	(By competitive supplier contract)	100
Ocean shipping	$2,300 per container × 6 containers	13,800
Marine insurance	($0.50 per hundred $)($32.67) (10,000)	1,663
Total CIF purchase price per month		$326,763
Total annual purchase price w/assists:	$326,763 × 12 = $3,921,156 + $6,000	$3,927,156
Additional costs per month:		
Port of entry (POE) terminal and handling	$700/container	$4,200
Custom's duties (4% of unit cost)	($32.67)(10,000)(4%)	13,068
Inland transportation-POE to facility	($20.00/100#)(15#/unit) (10,000)	30,000
Customs broker fees	(According to competitive contract)	200
Additional inventory carrying costs		
Warehousing, insurance, & taxes	($2/c.ft./month)(10,000 c. ft.) (.5 months)	10,000
Interest/investment capital foregone	($32.76)(10,000)((20%)	65,340
Bank fees for forward contract		1,000
Additional contract administration costs	5 hrs. × $30/hr.	150
Additional legal costs	16.34 hrs. × $200/hr.	3267
Additional communication costs	(10 long distance calls)	200
Total additional costs per month		$127,425
Total annual additional costs:		$1,529,100
Annual "total landed cost"		$5,456,256

$6,000. The total domestic cost then is $5,807,025. The buyer must ask the question: "Is the $350,769 differential sufficient justification for me to go off-shore?" If the domestic supplier is the "incumbent" supplier with which the firm has had a long-standing relationship, the answer to that question may be "no." If this is a new item for which there is no existing domestic supplier, the answer may be "yes." Many firms consider the degree of confidence in the estimated international sourcing costs, the firm's level of experience in off-shore sourcing compared with its level of experience in domestic sourcing, as well as its experience with a particular off-shore supplier in assessing whether a particular differential should dictate its decision. Many firms require a projected differential in order to accept an offer from an off-shore supplier. In the instant case, the difference between the domestic cost and the off-shore cost is only 6.0 percent. The buyer would need a high degree of confidence in the estimated total landed cost, in his or her ability to effect an efficient international purchase, and in the off-shore supplier's ability to deliver a high-quality product on time before deciding to go off-shore in the instant case.

Planning for Contingencies
Estimators commonly refer to some costs as "known-knowns," some costs as "known-unknowns," and other costs "unknown-unknowns." The costs considered to this point fit the known-knowns category (those for which we have firm quotations or factual information in hand), or known-unknowns (those for which we don't have firm quotations and hence must estimate). Other costs that could materialize in the future that we can't really estimate with any degree of certitude would fit the definition of unknown-unknowns. One of these unknown-unknowns is the increased risk of buyer-supplier disputes and the resultant costs of mediation, arbitration, or litigation.

There are potentially several different laws available applying to a typical international transaction. One of these laws is the Convention on Contracts for the International Sale of Goods (CISG). This convention is considered equivalent to (but different from) the Uniform Commercial Code, and can be made applicable to the transaction. Another law that can potentially be applied is the law of the exporter, while a third potential law is that of the importer (the law of the state where the importer conducts business). The contract terms must be very specific on which laws and which legal forum will apply. Different laws and forums can have different cost impacts on litigation and the amount of judgments made.

Many firms find that alternate dispute resolution (ADR) using some type of mediation or arbitration is preferred to petitioning a court of law. There are a plethora of rules and procedures available for use in international mediation and arbitration. Although the International Chamber of Commerce rules for arbitration are, perhaps, most well known, there are many others. The rules and how the body will be constituted may differ from organization to organization. Different rules and makeup can arrive at different consequences, some less favorable than others.

Those who believe they can avoid any mention of litigation in their off-shore contracts should be advised that such is not the case. Even if arbitration is upheld as a method of dispute resolution, there must be a legal forum available to interpret whether arbitration is applicable as well as to enforce the judgments of the arbitration body.

F.O.B. Origin Purchases

This situation is somewhat analogous to the international purchasing situation. The buyer will have available a price (in this case, the supplier's plant price) that does not represent the ultimate price the buyer will pay before the items are available for use. The buyer must apply to the supplier's plant price the costs of loading and unloading, transportation, and other associated costs. The buyer who fails to consider these costs in advance, and also fails to plan the solicitation and its required F.O.B. terms to arrive at the lowest ultimate overall cost for the purchase, will be doing his or her organization a great disservice.

PURCHASES FROM SUPPLIERS UNWILLING (OR UNABLE) TO PROVIDE COMPLETE COST DETAIL

Suppliers may resist submitting detailed cost information on a purchase, claiming such information is confidential and proprietary. In those all-too-familiar circumstances, the buyer will need to work with what he or she can get, either by negotiating with the supplier or obtaining the information from public sources. Some suppliers will give partial direct cost information without overhead, G & A, and profit rates. Others will provide major cost elements without elaboration on the makeup of those costs. Buyers confronted with this situation may compare the supplier's costs with known industry or market data or, if appropriate, with the cost accounting data available in the buyer's own organization (particularly where the buyer and supplier are in the same industry). If the supplier is a public company, the company's financial data will be available from public records. The buyer can use the earnings statement and other supporting schedules of expenses to derive estimates of the supplier's overhead G & A rates as well as determine the overall profit for the period being reviewed as measured against a number of bases. The supplier may be doing work for public clients, in which case the supplier's overhead and G & A rates will generally be reviewed and approved for prospective pricing purposes. Many suppliers will share such information with the buyer in order to foster and promote good business relationships and as a means of guaranteeing future work.

SUMMARY

The buyer will frequently be confronted with situations where price analysis will suffice for determining a fair and reasonable price for most of the value of the buy yet be insufficient for a portion of the total value. This situation is more common than most would admit. Purchasing situations calling for price analysis supplemented with partial cost analysis include purchases of modified commercial items, purchases from off-shore sources, F.O.B. Origin purchases, and purchases from suppliers unwilling (or unable) to provide complete cost detail.

9
Documenting the Cost and Price Analysis

ADEQUATE DOCUMENTATION

The buyer must document his or her actions. To be adequate, the record must show what was bought, why it was bought, how much competition was obtained, and why the price paid was considered reasonable. The purpose is to provide:

- Background to ensure informed decisions at each step in the purchasing process;
- Information for reviews and investigations conducted by internal review and higher management; and
- Essential facts in case of litigation.

Documentation can and should be simple. Each transaction must be documented. The buyer must consider the various written communications that support the contract pricing effort. The buyer must use established guidelines to record the results of price and cost analyses, describe the development of the negotiation objective, and summarize the results of the negotiation conference.

Records of analysis contribute to the factual basis for deciding that the offered price is fair and reasonable, and for setting the price objective if a negotiation conference is necessary. The memorandum of price negotiation and/or negotiation memorandum and any supporting documentation are used to support the proposed contract.

Moreover, due to personnel turnover, and use of contract files in succeeding purchases as well as in historical and investigative research, these reports and memoranda must permit a rapid reconstruction of major considerations of the particular pricing effort.

In summary, the official contract file must include a written document that demonstrates clearly that the price is reasonable. In making this demonstration, the document must show the significant facts that were considered in reaching agreement with the supplier. It also must indicate the extent to which the data submitted were relied on in establishing the prenegotiation objective and in reaching agreement.

The document, with its attachments, must identify the cost data used in the process. It must be complete enough in its narrative to show how the facts influenced judgments about future costs. It must show how the buyer got from contractor-furnished cost data, to the buyer's objective, and finally to the agreed-to price.

The amount of documentation needed depends on the type of action. If the contract award is made after price analysis, the documentation need not be as elaborate and detailed as for an award made after both price and cost analysis. This chapter covers recommended documentation for price analysis actions, and the more involved cost analysis documentation requirements.

DOCUMENTING PRICE ANALYSIS

The buyer documents in order to show that the contract price is fair and reasonable. This is accomplished by showing the price the supplier offered, the data used to evaluate the offer, the conclusions reached, and explaining why the conclusions were sound ones. Prepared forms, checklists, and spreadsheets aid in this process. Price analysis alone is the predominant method of analysis, particularly for low-value items. The buyer should document only to the extent necessary to ensure that the bottom line price is fair and reasonable.

DOCUMENTING COST ANALYSIS

The solicitation, where appropriate, must request offerors to submit cost proposals that will generally include cost data. An offeror is the primary source for its cost data, and the data are as much a part of the proposal as the price, delivery schedules, and other terms and conditions. These data include both supplier and subcontractor data, as appropriate. The solicitation should specify the kind and extent of pricing information the offeror must submit to support its proposal. The solicitation should instruct the offerors to use a standard estimating form (preferably the same one used to develop the in-house estimate) and to return a complete package so repeated requests for data and the necessary factual information will not be necessary. A "complete package" includes the following: a summary of all elements of cost and profit/fee, with references to exhibits supporting each element; a schedule of labor hours and rates by labor discipline, with extensions for each discipline and aggregated amounts for each labor category (engineering, manufacturing, service, and so forth); a consolidated, priced bill of materials, with explanations as to how materials were priced; budgeted one-year projections for overhead expense pools, bases, and rates, with all major accounting classifications shown (adjusted for known unallowables, as appropriate). In case of repeated negotiations with a company, an agreement with the company on the detail it will furnish with its proposals would be beneficial to both parties.

The buyer should request technical and accounting assistance, if available, and that request should generally be in writing. The buyer should request pricing assistance, if available, from all Pricing Team members as soon as possible in the procurement cycle. If the purchase has been programmed and constitutes an important part of the total work load, the buyer may want to line up the necessary pricing support for both supplier and subcontract proposals before the solicitation is prepared. Required data can be anticipated and requested in the solicitation so the necessary work can be scheduled on an orderly basis. In other instances, generally on smaller, more routine purchases, requests for pricing assistance can wait until the buyer has received the proposal and is satisfied that the proposal is complete and requires review.

Requests for assistance must be specific about the areas to be evaluated, the type of information wanted, and the time available for the analysis. The buyer should give the Pricing Team personnel as much time as possible. If there is particular concern with the estimated material cost, a request might be limited to an analysis of the cost of material. At the same time, a different specialist might be asked for some other information, such as an opinion on packaging costs.

When requesting information and assistance, the buyer should be extremely specific. To do this, the buyer must know something about the product or service being procured and its cost. Limiting the area of review will expedite the response. However, the specialist should be told the major areas of concern, the questions that need to be answered, and the report due date. How the specialist comes up with the answers is his or her speciality.

The report must convince the buyer that the specialist has done enough to ensure reliable answers, explored all reasonable avenues, and has not overlooked any significant factors. A request should never say: "Provide me an analysis of this entire proposal" unless nothing is known about the company or the problems that may be encountered in the analysis and negotiation.

In summary, the buyer should be as specific as possible about the scope of the review; limit the area of analysis whenever possible; set a due date for the reply; and recognize that while the request prescribes what the buyer wants, it should not tell the analyst how to do his or her job. Many proposals do not need complete analysis. Recent negotiations with the same company may have established a basis for negotiating the current proposal. The need in this case should be for only the latest, most current information on direct costs, with only limited verification of labor and indirect cost rates. In other cases, the proposal will be from a new company that could prompt a request for a complete analysis.

Requests for analysis of particular cost elements by engineers, packaging technicians, inspectors, and others also must be specific. Conserving their time is one, but not necessarily the most important, reason. The more specific the request, the more responsive will be the answer. It is not enough to get an answer in the jargon of the specialist; the buyer must be able to understand and use it.

Another basic ground rule is that the buyer is buying supplies, equipment, and services, *not costs*, and it is the ultimate lowest overall cost that buyer is concerned with. The buyer's objective in requesting pricing support is to have a factual basis for determining the value of what is being procured. By losing sight of this objective, it is easy to be caught up in the tangle of parochial quarrels over specific costs.

Technical reports are written by engineers or other technical specialists when requested by the buyer. They report findings that result from applying their special knowledge of the elements of the proposal. The auditor includes any financial effect of these findings in the report. Any differences between technical and audit analysis (such as projections of direct labor hours) are presented to the buyer.

The auditor (if involved in the action), prepares a report after reviewing the facts and evaluating the projections of future costs. This is done by reviewing the supplier's books and records. Since many companies won't permit buyer audit access, the buyer will often be required to rely on previous audit reports prepared by DOD or Federal Civilian Agency (DOE, EPA, DOC, and so forth) Auditors, state auditors, or independent auditors. The auditor will not generally be limited to those data submitted or identified in writing by the offeror. Moreover, if the audit was done previous to the proposal submission, the audit will not speak to the proposal at all. An auditor working on a specific proposal audit may comment on cost data not submitted by the offeror that may have a significant effect on the proposed contract price. Further, if the cost data submitted is not accurate, current, or complete, the auditor working on a specific proposal audit should identify the information in the audit report.

The buyer should use the advice and assistance of the Pricing Team in reaching a decision about the proposed price and in developing the negotiation objective. It is the responsibility of the buyer to reach an agreement with the offeror on a pricing arrangement.

For this reason, a buyer should generally not ask Pricing Team personnel to give a recommendation on the fairness and reasonableness of the contract as a whole or on total costs, proposed prices, or profit weights and rates. The buyer relies on these people for the facts on which to base the pricing decision and not for the decision itself. Generally, the buyer will know more about the overall situation and be in a better position to make these overall decisions. The buyer is looking for facts and informed opinions based on facts and generally asks the specialists for information on matters within their areas of special competence. A buyer should generally not ask the Pricing Team specialists for a negotiation objective, which would be a price recommendation.

The buyer should evaluate the quality of the pricing assistance and provide the sender a verdict whether good or bad, with supportable specifics. The senders should be aware of whether he or she has satisfied the needs as stated in the request for pricing assistance.

Written reports before negotiation generally should be restricted to pricing reports from the Pricing Team. Cost analysis reports, together with the offeror's cost data, are the other records used to form a prenegotiation objective. These are advisory reports, the means by which the person who made the analysis tells what was discovered. What the buyer does with this information is his or her own responsibility.

Ordinarily, conclusions must be expressed in terms of dollars or percentages if they are to be useful. An analyst who concludes that the offeror has overestimated the labor hours needed to do the job can't stop there. The advisory report also must state what estimated hours should be and why.

The buyer should include in his or her request for pricing support the prohibition that the Pricing Team member will not, unless specifically requested, make recommendations on subsequent actions or decisions. The buyer may, however, ask for a complete recommendation on price if deemed necessary and expedient.

If preliminary analysis leads to the conclusion that the offer is fair and reasonable and there is no need to hold a negotiation conference, a buyer report will be needed. It should summarize the actions taken and give the reasons why it is sound to award without further discussion. These instances occur infrequently.

Even though the buyer must develop an objective before entering negotiations with the company, it is not necessary to write the objective in a separate prenegotiation plan unless the purchase is for a significant amount.

If an auditor participates in analyzing the instant supplier's, the auditor's

150   Cost/Price Analysis: Tools to Improve Profit Margins

report should include a review of the influence of the contractor's actual costs on the estimate or proposal. The audit report should also identify the cost data reviewed.

Identification is not difficult; the offeror will have submitted or otherwise identified much of the cost required. A cost analysis report should identify only those additional or updated elements of data submitted in writing or otherwise acquired during the review process.

Cost analysis will require varying levels of detail or emphasis; length and detail will vary with the situation.

The buyer should collect and consolidate the several reports, adding comments and analyses as needed. If audit assistance was not requested, the buyer may be required to perform certain functions normally assigned to an auditor.

TYPICAL PRENEGOTIATION DOCUMENTATION

Prenegotiation cost price analysis contains two elements, both of which are addressed as follows:

1. Cost profit analysis, generally consisting of summary comparisons (spreadsheets reflecting the numbers) and cost-element by cost-element analysis (text rationale for the numerical positions).
2. Attachments, generally consisting of supplier and subcontractor cost proposal(s), technical analysis report(s) of supplier cost proposals, audit report(s) (if requested), and other Pricing Team report(s), as requested.

SUMMARY

The official contract file must include a written document that demonstrates clearly that the price is reasonable. In making this demonstration, the document must show the significant facts that were considered in reaching agreement with the supplier as well as the extent to which the data submitted were relied on in establishing the prenegotiation objective and in reaching agreement.

The amount of documentation needed depends on the type of action. If the contract award is made after price analysis, the documentation need not be as elaborate and detailed as for an award made after both price and cost analysis.

When documenting price analysis the buyer documents in order to show that the contract price is fair and reasonable. This is accomplished by showing the price the supplier offered, the data used to evaluate the offer, the conclusions reached, and explaining why the conclusions were sound. Pre-

prepared forms, checklists, and spreadsheets aid in this process. The buyer should document only to the extent necessary to convince himself or herself that the bottom line price is fair and reasonable.

When documenting cost analysis the buyer documents to reflect the cost proposals received from the suppliers, the different Pricing Team analyses of those proposals, and the prenegotiation position taken by the buyer based on those different analytical reports. Documenting cost analysis is generally more extensive than documentating price analysis.

10
Negotiating the Transaction

INTRODUCTION

This chapter provides guidelines for dealing with suppliers in negotiating prices or other matters. Although a full explanation of the negotiation process is beyond the scope of this manual, a few elementary points are necessary as a means of completing the process begun when the buyer requested a priced offer or cost proposal from the supplier.

Negotiation is the process of holding discussions with offerors to clarify any ambiguities or uncertainties, or to attempt to obtain better prices, terms or delivery, to discuss provisions of the proposed contract, and to examine any cost elements believed to be out of line. Negotiation can be very effective in improving value received and should be viewed as a positive opportunity to make a better deal. All purchasing personnel are encouraged to become skilled in negotiation practices and procedures and are encouraged to take one or more of the many specialized courses of instruction available for this purpose.

SITUATIONS WARRANTING NEGOTIATION

Negotiation may be accomplished informally by letters or telephone calls or formally through written correspondence or negotiation conferences, either at the buyer's or the supplier's facility. Regardless of the means used to communicate, every negotiation requires planning. A plan or list of discussion points is indispensable for conducting an effective negotiation. The following are illustrative of opportunities to negotiate a better transaction. Negotiation may be needed for a variety of reasons:

- The lowest quotation appears to be unreasonable.
- The buyer is purchasing a large number of items at one time.
- The item being acquired has a high mark-up or profit factor.
- The buyer does a large volume of business with the same supplier.
- There are alternative ways of furnishing the product or service.
- The low quoter has taken exception to the specifications or delivery conditions.
- The low quoter has taken exception to a required provision in the contract.
- The quotation can be made advantageous to the buyer with a minor improvement.
- The delivery time needs to be improved.
- For other justifiable reasons, including the need to award a contract on factors other than price.

Negotiate Only When Necessary

Negotiation for these and other reasons can be useful. It is necessary, however, to balance the potential benefit to be obtained against the administrative costs of negotiating. Complex negotiations are very rarely useful in most circumstances because they cost more to conduct than they can possibly save. In general, negotiations are generally conducted only with the source that has submitted the most favorable quotation. If those discussions do not lead to a satisfactory resolution of deficiencies, the next most favorable source should then be contacted.

CONDUCTING EFFECTIVE AND EFFICIENT NEGOTIATIONS

Negotiations should be conducted efficiently and quickly. In determining the extent of negotiations, the amount of the purchase should be considered. Because the dollar savings are potentially larger, more extensive negotiations are appropriate for higher dollar value orders.

Whatever the reason for holding discussions, the buyer should fully understand the nature of the requirement, the content of the solicitation, and the content of the quotations before beginning negotiations. All discussions should be positive and businesslike. The buyer should have sound reasons for requesting changes in a quotation. Merely indicating that a price is too high is a weak approach unless that claim can be supported with sound rationale. Remember however, that the buyer should never tell the quoter a particular price that must be met to gain further consideration. This is called the "auction" technique, and must be avoided. Key guidelines to conducting successful negotiations include the following:

- *Effective planning* is the key to successful negotiation. It is important to plan for negotiations by knowing the product or service and the market.
- *Flexibility* is the willingness to listen, evaluate, and rethink the buyer's initial position.
- *Mutual agreement as a goal* —try to find compromises that benefit both sides.
- *Never make promises that cannot be kept* —the buyer should be sure not to overstep his or her authority and should not make technical decisions without consulting technical personnel.
- *Seek assistance and counsel* from a more experienced buyer or technical personnel in both planning and conducting negotiations if necessary.
- *The negotiations for smaller dollar procurements* may only consist of asking for a better price, delivery, and so forth; however, in doing so other quoter prices should not be referred to or released.
- It is permissible to negotiate either with the lowest priced technically acceptable quoter or all technically acceptable quoters.

Following the discussions, the quoter must be allowed time to submit a revised quotation.

MAINTAINING FAIRNESS AND EQUITY AMONG SUPPLIERS

Throughout solicitation and negotiation, buyers and technical personnel must be careful not to take any action that might give one supplier an advantage over others. No information should be given to one supplier that is not given to all the other suppliers as well. This rule must be applied as follows:

- When suppliers call for clarification of requirements on a request for quotations, whether written or oral, no additional information can be given to one supplier that is not given to all. If clarification of an oral solicitation is necessary, the buyer must call each firm originally solicited, giving each the same information. If the solicitation was written, the buyer must send out an amendment to each recipient of the original solicitation.
- There can be no disclosure of either the number of quotations received or the content of any quotation before an order is placed.

- No information on an *anticipated* solicitation should be released in advance to a particular supplier. However, advance planning information may be made in a general announcement.

As a matter of professional courtesy, the buyer should notify unsuccessful suppliers. If an unsuccessful supplier requests an explanation of why it did not receive the award, a debriefing should be conducted in person or by phone. Normally, the explanation will be that the award was made to the lowest quoter. If this is not the case, the file should be clearly documented so that a good explanation may be given. The buyer should:

- Conduct the debriefing in such a way as to leave no doubt that the award decision was made fairly, impartially, and objectively.
- Take care never to disclose any proposer's or supplier's confidential business information such as prices, wage rates, and personnel.
- Refrain from discussing the relative positions of the unsuccessful suppliers.

NEGOTIATING CHANGE ORDERS

Because this is often a problematic area of purchasing, special attention should be given to negotiations surrounding changes to contracts. Generally speaking, contract modifications or change orders made to fixed-price contracts shall be fully priced and otherwise fully definitive at the time of issuance wherever possible. Authorizations to proceed (ATP) should be used only upon specific approval of the purchasing manager after proof of extreme urgency or other compelling reasons. Once approved, the ATP should be treated as an undefinitized subcontract modification, including an estimated cost, not to be exceeded, for the directed change or additional work. This estimate should not be in excess of funds available and include only work that can be accomplished prior to the target date for issuing a formal change order. In order to protect the buyer's interests, some form of cost or price analysis is required for all modifications, regardless of dollar value.

Unilateral change orders are those which are authorized by the "changes" clause of the contract. A typical changes clause provides that change orders can be issued unilaterally under specific situations without the consent of the supplier and without prior agreement as to price or time of performance.

Whenever possible, the buyer should attempt to obtain a firm quote on price and delivery from the supplier before issuing the change order. When the change order is issued without agreement on price and delivery, every effort should be made to reach agreement on the change as expeditiously as possible. The general guideline for definitizing change orders into modifications is to accomplish such action within 90 days after issuance.

The buyer should generally attain a firm not-to-exceed price prior to issuing the change order.

A typical changes clause requires the supplier to submit its claim for equitable adjustment within 30 days after receiving notice of the change, but that time period can be extended if the facts justify doing so.

Steps in Executing Contract Modifications

In executing contract modifications, the buyer should ensure the following:

1. Identify the contract clause under which the change occurred.
2. Develop a plan of action and milestones to achieve total, final pricing to include resolution of all costs attendant to the change, all issues of timeliness or delay, all disagreements on specs/drawings and quality control, and any other potential disagreement.
3. Organize information and document existing conditions before commencing changed work.
4. Define the limits and extent of changed work, including identification of deleted work.
5. Evaluate the necessity for change.
6. Obtain approval of the change as required.
7. Solicit a proposal from supplier (or acknowledge supplier's claim for change).
8. Prepare an in-house estimate.
9. Obtain reservation or commitment of funds for the proposed change. This may be based on the supplier's claim or not to exceed (NTE) and the in-house estimate.
10. Analyze the supplier's proposal. Request a quantitative analysis and audit if required.
11. Prepare, submit, or obtain approval of any client notification or approval requirements.
12. Negotiate total, final equitable adjustment.
13. Issue an ATP as appropriate.
14. Prepare the negotiation memorandum.
15. Submit the contract file for any required review and approval.
16. Issue a formal contract modification (change order).

Procedures for Issuing an ATP

In the event it becomes necessary to issue an ATP, such written directive should be processed in accordance with the following:

1. The buyer should proceed as indicated in 1 through 6, as well as 8 and 9 in the preceeding paragraph.
2. Upon receipt of funds, official approval, and in-house estimate, the buyer should prepare an ATP in accordance with established purchasing policy. The target date established for completing negotiations of a formal change order (normally 30 days from date of the ATP) should be met. It is imperative that negotiations be initiated immediately after the ATP letter has been issued and pursued diligently.
3. The buyer should periodically report to purchasing management the status of all undefinitized ATPs.

If agreement cannot be reached on an equitable adjustment for a unilateral change order, the buyer should escalate the matter to purchasing management for disposition.

SUMMARY

Negotiation is the process of holding discussions with offerors to clarify any ambiguities or uncertainties, or to attempt to obtain better prices, terms or delivery, to discuss provisions of the proposed contract, and to examine any cost elements believed to be out of line. Negotiation can be very effective in improving buyer value received, and should be viewed as a positive opportunity to make a better deal. All purchasing personnel are encouraged to become skilled in negotiation practices and procedures and are encouraged to take one or more of the many specialized courses of instruction available for this purpose.

Negotiation may be accomplished informally by letters or telephone calls or formally through written correspondence or negotiation conferences, either at the buyer's or the supplier's facility. Regardless of the means used to communicate, every negotiation requires planning. A plan or list of discussion points is indispensable for conducting an effective negotiation. Negotiation may be needed because the lowest quotation appears to be unreasonable; the buyer is purchasing a large number of items at one time; the item being acquired has a high mark-up or profit factor; the buyer does a large volume of business with the same supplier; there are alternative ways of furnishing the product or service; the low quoter has taken exception to the specifications or delivery conditions; the low quoter has taken exception to a required provision in the contract; the quotation can be made advantageous to the buyer with a minor improvement; the delivery time needs to be improved; or for other justifiable reasons, including the need to award a contract on factors other than price.

Key guidelines to conducting successful negotiations include effective planning; flexibility; mutual agreement; refraining from making promises that cannot be kept; seeking assistance and counsel when necessary; and keeping negotiations streamlined and simple.

Special attention should be given to negotiations surrounding changes to contracts. Generally speaking, contract modifications or change orders made to fixed price contracts should be fully priced and otherwise fully definitive at the time of issuance wherever possible. Authorizations to proceed should be used only upon specific approval of the purchasing manager after proof of extreme urgency or other compelling reasons. Once approved, the ATP should be treated as an undefinitized subcontract modification, including an estimated cost, not to be exceeded, for the directed change or additional work. This estimate should not be in excess of funds available and include only work that can be accomplished prior to the target date for issuing a formal change order. The buyer's objective in change order negotiation should be to negotiate on a prospective basis as much of the cost of performance as possible. This minimizes the buying organization's exposure and forces the supplier to be efficient. Some form of cost or price analysis is required for all modifications, regardless of dollar value.

11
Documenting the Negotiation

NEGOTIATION MEMORANDUM

A negotiation memorandum will generally be required to document the negotiation. This document will generally be the responsibility of the buyer (except, in those rare instances where the buyer has delegated the responsibility for negotiation). The negotiation memorandum should be the only document required after negotiations have been concluded. It tells the reader the story of the negotiations. What were the offered prices and what were the offered costs? What was the buyer's price objectives and what were the costs supporting that goal? What costs were submitted but not relied on and not used? What were the goals for delivery and pricing arrangement? What was discussed? What were the compelling arguments? What disposition was made of the principal points raised in preliminary analyses; included in the objective, and discussed in the negotiations? What values, cost, and other factors support the agreed-to price? If these are different from those supporting the objective, what justifications are there for the difference?

The negotiation memorandum is, first, a sales document that establishes the reasonableness of the agreement reached with the supplier. Second, it is the permanent record of the buyer's decision. It charts the progress from proposal through negotiations and does so in specifics.

The negotiation memorandum will be the source document if it becomes necessary to reconstruct the events of the purchase. The buyer may not be around to help, so he or she must leave tracks that strangers can follow. In addition to proving that the price is fair and reasonable, the negotiation memorandum must identify data relied on and that data submitted but not relied on. It should convince the reader that the buyer and Pricing Team did all that needed to be and could be done to reach a fair and reasonable price.

Ultimately, the buyer will find himself or herself writing for the most important of probable readers: the individual or group that has the final say on whether the contract is approved. This will dictate the detail of the memorandum and, to some extent, its style. The buyer should strike a balance between too few words and too many.

Format for Negotiation Memorandum

The purchasing manager should establish a standard format for negotiation memoranda. While the format is standard, the content must vary to report the actual events of the analysis and negotiation. What events will be reported depends on whether the negotiation is to agree on the terms of a definitive contract, a definitive contract superseding a letter contract, new work added to an existing contract, or the final settlement of a termination.

Every negotiation memorandum should contain a postnegotiation summary. This segment should show the supplier's price or cost proposal, the buyer's negotiation objective, and the negotiation results tabulated in parallel form. If the supplier has submitted a cost proposal that has been analyzed by the buyer, there should be a side-by-side comparison of major elements of cost and profit. Whether these will be summary figures for total contract value, summary for the total price of the major item, unit price for the major item, or some other presentation will depend on how the negotiations were conducted. The general rule is to portray the negotiation as it actually took place. Unit cost and profit figures may not give the true picture of the significance of each element; the buyer should show total as well as unit values in the narrative that follows. A sample format for this breakdown is shown in Appendix C.

For prospective price negotiations, estimated profit is an integral part of the negotiation objective. The development of the profit objective for those negotiations should be discussed in this part of the negotiation memorandum.

A typical negotiation memorandum would contain the following basic elements:

1. Results of negotiation, generally consisting of summary comparisons, a discussion of reasons for differences between prenegotiation and final negotiated positions, an identification of final cost data and the degree to

which the data was relied upon, and an identification of supplemental data obtained during negotiation and an evaluation of the data.
2. Attachments, including supplier (and subcontractor, if appropriate) cost proposals submitted during negotiation, with best and final cost proposals and incentive share arrangements (if any).

SUMMARY

A negotiation memorandum will generally be required to document the negotiation. This document will generally be the responsibility of the buyer. The negotiation memorandum should be the only document required after negotiations have been concluded. It tells the reader the story of the negotiations.

The negotiation memorandum is, first, a sales document that establishes the reasonableness of the agreement reached with the supplier. Second, it is the permanent record of the buyer's decision. It charts the progress from proposal through negotiations and does so in specifics.

The negotiation memorandum will be the source document if it becomes necessary to reconstruct the events of the purchase. The buyer may not be around to help, so he or she must leave tracks that strangers can follow. In addition to proving that the price is fair and reasonable, the negotiation memorandum must identify data relied on and that data submitted but not relied on. It should convince the reader that the buyer and Pricing Team did all that needed to be and could be done to reach a fair and reasonable price.

Ultimately, the buyer will find himself or herself writing for the most important of probable readers: the individual or group that has the final say on whether the contract is approved. This will dictate the detail of the memorandum and, to some extent, its style. The buyer should strike a balance between too few words and too many.

The purchasing manager should establish a standard format for negotiation memoranda. While the format is standard, the content must vary to report the actual events of the analysis and negotiation.

Every negotiation memorandum should contain a postnegotiation summary. This segment should show the supplier's price or cost proposal, the buyer's negotiation objective, and the negotiation results tabulated in parallel form.

12

"Strategic Cost Analysis" Techniques Available to the Purchasing Manager

DEFINITION

Although analyzing individual purchase transactions can more than pay for the effort required, "strategic cost analysis" has the potential to provide rich cost-savings rewards to the purchasing manager and his or her organization. Recall we defined strategic cost analysis as those broad-based, organization-wide efforts designed to effect purchasing savings in the long-run.

CIRCUMSTANCES WHERE STRATEGIC COST ANALYSIS IS APPROPRIATE

Included in this type of analysis would be the following major categories, listed in a general purchasing cycle order or chronology:

- Changes in quality or design of the products and services purchased.
- Changes in delivery and order quantity requirements.
- Changes in supplier base.
- Changes in procurement and buying methods.
- Changes in inventory and materials management.

The first two of these categories relate to the requirements-generation process. In these types of analysis, the purchasing manager must work closely with users of the supplies or services being purchased to assure they consider the various cost-savings techniques available. With respect to these first two categories, the purchasing manager must place himself or herself in a supporting role. In the next two types of analysis (supplier base and procurement or buying methods), the purchasing manager takes the lead, with vigorous support from requirements and other materials functions. With respect to the last of the types of analysis, the purchasing manager supports the inventory manager (except in those obvious instances where the organization has placed the inventory management function in purchasing).

Changes in Quality or Design of the Products or Services Purchased

This is a very fertile ground for cost-savings. Potential redesign efforts should address not only the materials and components used in manufacturing the final product or service but also the manufacturing processes themselves and the incoming and outgoing packaging of the materials and products. Potential cost-saving efforts could include specifying lower grade material or more durable material, using standardized or simpler materials, standardizing simplifying the manufacturing processes used, or redesigning the packaging containers or even the labeling system for the containers.

Changes in Delivery and Order Quantity Requirements

This activity is second only to quality and design changes in offering substantial cost savings. By analyzing bottom line cost or landed cost of materials, the user organization (assisted by purchasing) can favorably impact the cost of purchasing. Potential cost savings could include packaging material requirements so as to deliver in larger quantities (consistent with whatever JIT and inventory minimization goals the organization has); use of make-or-buy analysis of the transportation component of the incoming materials prices in order to get the best bang for the buck; and changing modes of transportation (truck to rail or rail to barge/boat, and so forth). When considering packaging requirements, the user organization should consider using local off-site storage, increasing storage capacity in-house, or using supplier storage capacity.

Changes in Supplier Base

This is where the purchasing manager can shine. Supplier base management should present numerous opportunities for savings. By carefully analyzing production and MRO materials purchased by the organization, the purchasing manager can decide whether to expand or contract the supplier base for specific materials and services, change suppliers because of quality or other deficiencies, or even relocate certain operations in order to effect closer integration with supplier partners.

Changes in Procurement and Buying Methods

This is another area where the purchasing manager should be able to work magic. In this area, the purchasing manager should try to think strategically. This type of broad thinking may discover opportunities for the company to practice "vertical integration." Rather than partnering with a supplier, the organization may find it beneficial to acquire the supplier's operation, either through outright purchase or through effective financial control. Another broad-based method goes to the heart of the source-selection decision: "Should an individual item be purchased or produced in-house?" A very aggressive make-or-buy program can pay most organizations big dividends. Very closely allied with make-or-buy is the lease/purchase decision process. This type of economic/financial analysis can uncover hidden opportunities to effect cost savings through the simple decision of whether to own a particular item of equipment or property. Less broad sweeping, but often equally effective, is using long-term contracts, systems contracts, requirements contracts, and indefinite-quantity/indefinite-delivery contracts.

Changes in Inventory and Materials Management

Although these activities come later in the purchasing cycle after most of the more potent decisions have been made, these cost-savings measures should not be overlooked by the organization. Potential cost-saving efforts include all the various ways the inventory manager can reduce inventory investment and improve management procedures, including use of ABC analysis, use (and frequent update) of EOQ procedures, reduction in safety stock, efficient and effective use or disposition of obsolete stock, and use of computerized perpetual inventory procedures, if appropriate.

The purchasing manager who proves adept at these more general, non-transaction-specific analytical methods will be able to take great personal and professional satisfaction in having made a significant contribution to the effectiveness and efficiency of his or her organization.

SUMMARY

Although analyzing individual purchase transactions can more than pay for the effort required, strategic cost analysis has the potential to provide rich cost-savings rewards to the purchasing manager and his or her organization. Strategic cost analysis are those broad-based, organizationwide efforts designed to effect purchasing savings in the long-run. Included in this type of analysis are changes in quality or design of the products and services purchased, changes in delivery and order quantity requirements, changes in supplier base, changes in procurement and buying methods, and changes in inventory and materials management. All of these efforts can pay back much larger dividends than the transaction-based analytical methods discussed in previous chapters.

Glossary

Adequate price competition exists when offers are solicited and two or more responsible offerors who can satisfy the solicitation requirements submit offers to the solicitation's expressed requirements, and these offerors compete independently for a contract to be awarded to the responsible offeror submitting the lowest evaluated price. Price must be a substantial factor in the evaluation, and the award should generally be based on price analysis alone. A price may be justified based on adequate price competition if it results directly from price competition or if price analysis alone clearly demonstrates that the proposed price is reasonable in comparison with current or recent prices for the same or substantially the same items purchased in comparable quantities, terms, and conditions under contracts that resulted from adequate price competition.

Allocable cost is assignable or chargeable to one or more specific cost objectives in accordance with the relative benefits received or other equitable relationships defined or agreed to between contractual parties.

Allowable cost meets the tests of reasonableness and allocability, conforms to generally accepted accounting principles, or agreed-upon terms between the contractual parties.

Cost analysis is the review and evaluation of the separate cost elements included in an offeror's cost proposal, including the judgmental factors applied by the offeror in projecting from historical cost data to the estimated costs included in the cost proposal, in order to form an opinion on the degree to which the proposed costs represent what the contract should cost, assuming reasonable economy and efficiency. It includes the verification of cost data, and evaluation of cost elements, including:

- The necessity for and reasonableness of proposed costs.
- Projection of the offeror's cost trends, on the basis of current and historical cost data.
- A technical appraisal of the estimated labor, material, tooling and facilities requirements, and of the reasonableness of scrap and spoilage factors.
- The application of approved indirect cost rates, labor rates, or other factors.

Among the evaluations that should be made, where the necessary data are available, are comparisons of an offeror's current estimated costs with:

- actual costs previously incurred by the same supplier or offeror;
- previous cost estimates from the offeror or from other offerors for the same or similar items;
- other cost estimates received in response to the solicitation;
- independent costs estimates by technical personnel; and
- forecasts of planned expenditures.

Cost-reimbursement contracts refers to a family of pricing arrangements or contract types that provide for payment of allowable, allocable, and reasonable costs incurred in the performance of a contract, to the extent that such costs are prescribed or permitted by the contract. These contracts establish an estimate of total cost for the purpose of obligating funds and establishing a ceiling that the contractor may not exceed (except at its own risk) without the approval of the buyer. Types of cost-reimbursement contracts include: (1) cost, (2) cost-sharing, (3) cost-plus-incentive-fee, (4) cost-plus-award-fee, and (5) cost-plus-fixed-fee.

Direct cost is a cost that can be assignable to specific products, usually classified as direct labor cost, direct material cost, or purchased cost. These costs are usually treated as variable and do not include overhead or common cost allocations.

Established catalog prices are prices included in a catalog, price list, schedule, or other form that is regularly maintained by the manufacturer or supplier, is either published or otherwise available for inspection by customers; and states prices at which sales are currently being made to a significant number of buyers constituting the general public. In order to use catalog prices for price analysis, the item must be a commercial item sold in substantial quantities to the general public. It includes items sufficiently similar to such commercial items to permit the differences in prices to be identified and justified without resort to cost analysis.

Established market prices are current prices established in the course of ordinary and usual trade between buyers and sellers free to bargain that can be substantiated by data independent of the manufacturer or supplier.

Fee is a term used in lieu of profit to refer to the payment for risk in a cost-reimbursement contract arrangement. The fee may be fixed initially (cost-plus-fixed-fee), or it may vary (cost-plus-incentive-fee or cost-plus-award-fee).

Fixed cost is a cost that is primarily related to a given time period and does not change due to production volume. Examples include rent, depreciation, and property taxes.

Fixed-price contracts refers to a family of pricing arrangements or contract types whose common discipline is a ceiling beyond which the buyer bears no responsibility for payment. Types include: (1) firm-fixed-price, (2) fixed-price with economic price adjustment/escalation, (3) fixed-price incentive, (4) fixed-price with prospective price redetermination (rarely used), (5) fixed-price with retroactive price redetermination (also rarely used), and (6) firm-fixed-price, level of effort term contracts (also rarely used).

Indirect cost is a cost not directly identified with a single final cost objective; also known as overhead.

Price analysis is the process of examining and evaluating a proposed price without evaluating its separate cost elements and profit. It may be accomplished by the following comparisons, which are listed in their relative order of preference:

- Comparison with other prices and quotations submitted.
- Comparison with published catalog or market prices.
- Comparison with prices set by law or regulation.
- Comparison with prices for the same or similar items.
- Comparison with prior quotations for the same or similar items.
- Comparison with market data (indexes).
- Application of rough yardsticks (such as dollars per pound or per horsepower or other units) to highlight significant inconsistencies that warrant additional pricing inquiry.
- Comparison with independent estimates of cost developed by knowledgeable personnel within the buying organization.
- Use of value analysis.
- Use of visual analysis.

Procurement methods refer to procedures used in the preaward phase to arrive at a source-selection decision preparatory to award. The method used generally determines the evaluation, process, pricing approach, and type of contract used. There are generally two procurement methods, with several modifications of each. The first of these (the simpler of the two) is the price competitive approach. This is accomplished in the private sector by use of competitive bidding. In the public sector, sealed bidding may be required. Both competitive bidding and sealed bidding rely heavily on award to the lowest bidder and generally result in a fixed-price contract. The second method (more complex than the first), is other than price competitive, often employing selection based on factors in addition to price, extensive cost analysis, and technical-cost negotiations. In this second method, other than fixed-price contracts are some-

172 Cost/Price Analysis: Tools to Improve Profit Margins

    times awarded. Procurement methods should not be confused with ordering methods and payment methods.

Profit is generally characterized as the basic motive of business enterprise, on occasion referred to as "the wages of risk." In contract pricing, profit represents a projected or known monetary excess realized by a supplier after the deducting cost (both direct and indirect) incurred or to be incurred in performing job, task, or series of the same.

Prospective pricing refers to a pricing decision made in advance of performance, based on analyzing comparative prices, cost estimates, past costs, or combinations of such considerations.

Reasonableness is a test given to costs to determine if their nature or amount does not exceed what would be incurred by an ordinarily prudent person in the conduct of competitive business.

Retroactive pricing refers to a pricing decision made after some or all of the work specified under contract has been completed, based on a review of performance and recorded cost data.

Single source is characterized as one source among others in a competitive marketplace that, for justifiable reasons, is found to be most advantageous for the purpose of contract award or is the only source quoting on a given purchase.

Sole source is characterized as the one and only source regardless of the marketplace, possessing a unique and singularly available performance capability for the purpose of contract award.

Strategic cost analysis is characterized as broad-based, organizationwide efforts designed to effect purchasing savings in the long-run. This type of analysis is not related to specific purchasing transactions.

Transactional analysis is characterized as narrowly based, individual buyer (with expert assistance, as appropriate) efforts designed to effect purchasing savings in the instant purchase order or contract.

Variable cost is a cost that fluctuates with the rate of production of goods or the performance of services.

Reference List

Amembal, Sudhir P. *Equipment Acquisition and the Lease Financing Alternative.* NAPM International Purchasing Conference Proceedings. Tempe, Arizona: National Association of Purchasing Management, 1990.

Anklesaria, Jimmy. *Cost Management Strategies for the 1990s.* NAPM International Purchasing Conference Proceedings. Tempe, Arizona: National Association of Purchasing Management, 1992.

Anklesaria, Jimmy. *Zero Base Pricing: Achieving Competitiveness Through Reduced All-in-Cost.* NAPM International Purchasing Conference Proceedings. Tempe, Arizona: National Association of Purchasing Management, 1990.

Benson, Gregory E. *Anything is Negotiable!* NAPM International Purchasing Conference Proceedings. Tempe, Arizona: National Association of Purchasing Management, 1990.

Benson, Gregory E. *A Positive Approach to Negotiations.* NAPM International Purchasing Conference Proceedings. Tempe, Arizona: National Association of Purchasing Management, 1992.

Brusman, Calvin. *Best Value Source Selection—Made Easy.* NAPM International Purchasing Conference Proceedings. Tempe, Arizona: National Association of Purchasing Management, 1992.

Bonneville Power Administration. *Guide to Preparation of Documentation for Cost/Price Analysis and Negotiation of Contracts.* Portland, Oregon: Bonneville Power Administration, 1987.

Bureau of Labor Statistics, U.S. Department of Labor. *Consumer Price Index Detailed Report.* Washington: U.S. Government Printing Office, 1992.

Bureau of Labor Statistics, U.S. Department of Labor. *Producer Prices and Price Indexes.* Washington: U.S. Government Printing Office, 1992.

Burt, David N. *How to Implement Zero Base Pricing.* NAPM International Purchasing Conference Proceedings. Tempe, Arizona: National Association of Purchasing Management, 1990.

Burt, David N. *The Zero Base Pricing: Approach to Purchasing Consulting Services.* NAPM International Purchasing Conference Proceedings. Tempe, Arizona: National Association of Purchasing Management, 1990.

Cashin, James A. *Cashin's Handbook for Auditors*, 2nd ed. New York: McGraw-Hill, 1986.

Cleland, David. *Project Management Handbook.* New York: Van Nostrand Reinhold, 1988.

Doak, Marshall J. Jr. *Cost or Pricing Data—An End to the Fact Versus Judgment Confusion.* Federal Contracts Report, April 14, 1992. Washington, D.C.: The Bureau of National Affairs, Inc., 1992.

Dobler, Donald W., David N. Burt, and Lamar Lee, Jr.. *Purchasing and Materials Management*, 5th ed. New York: McGraw-Hill, 1990.

Ellram, Lisa M. *Selling Purchasing Cost Savings: Tools and Trends.* NAPM International Purchasing Conference Proceedings. Tempe, Arizona: National Association of Purchasing Management, 1991.

Elsayed, Elsayed A. *Analysis and Control of Production Systems.* Englewood Cliffs, NJ: Prentice-Hall, 1985.

Fulford, Dennis L. *Taking the Offensive in Cost Control.* NAPM International Purchasing Conference Proceedings. Tempe, Arizona: National Association of Purchasing Management, 1989.

Glenn, Patti and Robert Shealy. *Understanding Total Cost of Ownership: A Case Study Approach.* NAPM International Purchasing Conference Proceedings. Tempe, Arizona: National Association of Purchasing Management, 1989.

Graw, LeRoy H. *Cost, Price, and Profit Analysis Techniques.* NAPM International Purchasing Conference Proceedings. Tempe, Arizona: National Association of Purchasing Management, 1990.

Graw, LeRoy H. *RFP Process and Product Simplification.* NAPM International Purchasing Conference Proceedings. Tempe, Arizona: National Association of Purchasing Management, 1992.

Horngren, Charles T. *Cost Accounting: A Managerial Emphasis*, 7th ed. New York: Prentice-Hall, 1991.

Huard, Mary Jo. *Winning Negotiation Strategies.* NAPM International Purchasing Conference Proceedings. Tempe, Arizona: National Association of Purchasing Management, 1990.

Janson, Robert L. *Least Total Cost Purchasing.* NAPM International Purchasing Conference Proceedings. Tempe, Arizona: National Association of Purchasing Management, 1990.

John, Pamela and Debra Sisk. *Risk Management Techniques for the Purchasing Manager.* NAPM International Purchasing Conference Proceedings. Tempe, Arizona: National Association of Purchasing Management, 1991.

Kemp, Robert A. *Organizing to Control Costs in International Purchasing Operations.* NAPM International Purchasing Conference Proceedings. Tempe, Arizona: National Association of Purchasing Management, 1989.

Laske, Walter F. *The Total Cost of Purchasing Abroad: "Hidden Costs."* NAPM International Purchasing Conference Proceedings. Tempe, Arizona: National Association of Purchasing Management, 1989.

Long, Brian G. *What Purchasers Should Know About Economics.* NAPM International Purchasing Conference Proceedings. Tempe, Arizona: National Association of Purchasing Management, 1991.

Marshall, A. *More Profitable Pricing.* New York: McGraw-Hill, 1980.

Melvin, Michael. *International Money and Finance*, 2nd ed. New York: Harper Collins, 1989.

Monczka, Robert M. and Larry C. Giunipero. *Purchasing Internationally: Concepts and Principles.* Chelsea, Michigan: BookCrafters, 1990.

National Association of Purchasing Management, Inc. *NAPM Insights.* Tempe, Arizona: National Association of Purchasing Management, Inc., 1992.

National Contract Management Association. *Contract Management.* Vienna, Virginia: National Contract Management Association, 1992.

Norquist, Warren E. *Zero Base Pricing: Achieving Competitiveness through Reduced All-in-Cost.* NAPM International Purchasing Conference Proceedings. Tempe, Arizona: National Association of Purchasing Management, 1990.

Plossl, George W. *Production and Inventory Control—Applications*, 2nd ed. Atlanta, Georgia: George Plossl Educational Services, Inc, 1985.

Pooler, Victor H. *Reaching for the World of Global Purchasing.* New York: Van Nostrand Reinhold, 1992.

Sanchez, S.J. and W.J. Bischoff. *Buyer-Seller Triangle.* NAPM International Purchasing Conference Proceedings. Tempe, Arizona: National Association of Purchasing Management, 1990.

Salvendy, Gavriel. *Handbook of Industrial Engineering.* New York: John Wiley, 1991.

Santos, Denise L. *Do's and Don'ts for Successful Negotiations.* NAPM International Purchasing Conference Proceedings. Tempe, Arizona: National Association of Purchasing Management, 1990.

U.S. Air Force Institute of Technology, School of Systems and Logistics. *Principles of Contract Pricing, 4th ed.* Wright Patterson Air Force Base, Ohio, Undated.

U.S. Army Missile Command. *Alpha & Omega and the Experience Curve.* Redstone Arsenal, Alabama: U.S. Army Missile Command, Undated.

U.S. Department of Defense. *Armed Services Pricing Manual, 1986, Volume 1, Contract Pricing.* Washington: Government Printing Office, 1986.

U.S. Department of Defense. *Armed Services Pricing Manual, 1987, Volume 2, Price Analysis.* Washington: Government Printing Office, 1987.

U.S. Department of Defense, General Services Administration, National Aeronautics and Space Administration. *Federal Acquisition Regulation* (with amendments through May 1992). Washington: Government Printing Office, 1984.

U.S. Department of Defense, Defense Contract Audit Agency. *DCAA Contract Audit Manual.* Washington: Government Printing Office, 1992.

Watson, D.S. *Price Theory and Its Uses.* Lanham MD: University Press of America, 1991.

Wilson, John D. *Total Cost Vs. Price—Understanding the Difference.* NAPM International Purchasing Conference Proceedings. Tempe, Arizona: National Association of Purchasing Management, 1990.

Appendix A

Tables of Learning Curve Data

LEARNING CURVE

Percent Learning:	75.00%			76.00%		
N	CUM TOTAL	CUM AVG	UNIT	CUM TOTAL	CUM AVG	UNIT
1	1.00000000	1.00000000	1.00000000	1.00000000	1.00000000	1.00000000
2	1.75000000	.87500000	.75000000	1.76000000	.88000000	.76000000
3	2.38383583	.79461194	.63383583	2.40728272	.80242757	.64728272
4	2.94633583	.73658396	.56250000	2.98488272	.74622068	.57760000
5	3.45908060	.69181612	.51274477	3.51364167	.70272833	.52875895
6	3.93445748	.65574291	.47537687	4.00557654	.66759609	.49193487
7	4.38037303	.62576758	.44591556	4.46838517	.63834074	.46280863
8	4.80224803	.60028100	.42187500	4.90736117	.61342015	.43897600
9	5.20399589	.57822177	.40174786	5.32633609	.59181512	.41897492
10	5.58855447	.55885545	.38455858	5.72819289	.57281929	.40185680
11	5.95819789	.54165435	.36964342	6.11516777	.55592434	.38697488
12	6.31473055	.52622755	.35653266	6.48903827	.54075319	.37387050
13	6.65961351	.51227796	.34488296	6.85124614	.52701893	.36220787
14	6.99405018	.49957501	.33443667	7.20298070	.51449862	.35173456
15	7.31904618	.48793641	.32499601	7.54523723	.50301582	.34225653
16	7.63545243	.47721578	.31640625	7.87885899	.49242869	.33362176
17	7.94399675	.46729393	.30854432	8.20456816	.48262166	.32570917
18	8.24530765	.45807265	.30131090	8.52298910	.47349939	.31842094
19	8.53993245	.44947013	.29462480	8.83466612	.46498243	.31167702
20	8.82835138	.44141757	.28841893	9.14007729	.45700386	.30541117
21	9.11098864	.43385660	.28263726	9.43964532	.44950692	.29956803
22	9.38822121	.42673733	.27723257	9.73374623	.44244301	.29410091
23	9.66038596	.42001678	.27216476	10.02271632	.43577027	.28897009
24	9.92778546	.41365773	.26739949	10.30685790	.42945241	.28414158
25	10.19069265	.40762771	.26290720	10.58644393	.42345776	.27958603
26	10.44935487	.40189826	.25866222	10.86172191	.41775854	.27527798
27	10.70399706	.39644434	.25464219	11.13291714	.41233026	.27119523
28	10.95482456	.39124373	.25082750	11.40023540	.40715126	.26731827
29	11.20202543	.38627674	.24720087	11.66386532	.40220225	.26362992
30	11.44577244	.38152575	.24374701	11.92398029	.39746601	.26011496
31	11.68622476	.37697499	.24045232	12.18074015	.39292710	.25675986
32	11.92352944	.37261030	.23730469	12.43429269	.38857165	.25355254
33	12.15782269	.36841887	.23429325	12.68477484	.38438712	.25048215
34	12.38923093	.36438915	.23140824	12.93231381	.38036217	.24753897
35	12.61787180	.36051062	.22864087	13.17702802	.37648651	.24471421
36	12.84385497	.35677375	.22598317	13.41902793	.37275078	.24199991
37	13.06728291	.35316981	.22342794	13.65841681	.36914640	.23938888
38	13.28825151	.34969083	.22096860	13.89529134	.36566556	.23687453
39	13.50685069	.34632950	.21859918	14.12974224	.36230108	.23445090
40	13.72316489	.34307912	.21631420	14.36185473	.35904637	.23211249
41	13.93727354	.33993350	.21410865	14.59170902	.35589534	.22985429
42	14.14925149	.33688694	.21197794	14.81938072	.35284240	.22767170
43	14.35916932	.33393417	.20991783	15.04494120	.34988235	.22556047
44	14.56709375	.33107031	.20792443	15.26845789	.34701041	.22351669
45	14.77308786	.32829084	.20599411	15.48999463	.34422210	.22153674
46	14.97721143	.32559155	.20412357	15.70961190	.34151330	.21961727
47	15.17952112	.32296853	.20230969	15.92736708	.33888015	.21775519
48	15.38007074	.32041814	.20054962	16.14331468	.33631906	.21594760
49	15.57891142	.31793697	.19884068	16.35750652	.33382666	.21419183
50	15.77609182	.31552184	.19718040	16.56999190	.33139984	.21248538

LEARNING CURVE

Percent Learning:	75.00%			76.00%		
N	CUM TOTAL	CUM AVG	UNIT	CUM TOTAL	CUM AVG	UNIT
51	15.97165826	.31316977	.19556645	16.78081781	.32903564	.21082592
52	16.16565493	.31087798	.19399666	16.99002908	.32673133	.20921127
53	16.35812396	.30864385	.19246903	17.19766847	.32448431	.20763939
54	16.54910560	.30646492	.19098164	17.40377684	.32229216	.20610837
55	16.73863833	.30433888	.18953273	17.60839327	.32015260	.20461643
56	16.92675896	.30226355	.18812063	17.81155515	.31806348	.20316188
57	17.11350271	.30023689	.18674376	18.01329830	.31602278	.20174315
58	17.29890337	.29825695	.18540065	18.21365704	.31402857	.20035874
59	17.48299329	.29632192	.18408992	18.41266429	.31207906	.19900725
60	17.66580354	.29443006	.18281025	18.61035166	.31017253	.19768737
61	17.84736396	.29257974	.18156042	18.80674951	.30830737	.19639785
62	18.02770320	.29076941	.18033924	19.00188701	.30648205	.19513750
63	18.20684882	.28899760	.17914562	19.19579222	.30469511	.19390521
64	18.38482733	.28726293	.17797852	19.38849215	.30294519	.19269993
65	18.56166427	.28556407	.17683693	19.58001280	.30123097	.19152065
66	18.73738420	.28389976	.17571994	19.77037923	.29955120	.19036644
67	18.91201084	.28226882	.17462663	19.95961561	.29790471	.18923638
68	19.08556702	.28067010	.17355618	20.14774523	.29629037	.18812962
69	19.25807479	.27910253	.17250777	20.33479058	.29470711	.18704535
70	19.42955544	.27756508	.17148065	20.52077337	.29315391	.18598280
71	19.60002952	.27605675	.17047408	20.70571460	.29162978	.18494123
72	19.76951690	.27457662	.16948738	20.88963453	.29013381	.18391993
73	19.93803678	.27312379	.16851988	21.07255279	.28866511	.18291825
74	20.10560774	.27169740	.16757095	21.25448833	.28722282	.18193555
75	20.27224774	.27029664	.16664000	21.43545954	.28580613	.18097120
76	20.43797419	.26892071	.16572645	21.61548418	.28441427	.18002465
77	20.60280394	.26756888	.16482975	21.79457950	.28304649	.17909531
78	20.76675333	.26624043	.16394938	21.97276218	.28170208	.17818268
79	20.92983817	.26493466	.16308484	22.15004841	.28038036	.17728624
80	21.09207382	.26365092	.16223565	22.32645391	.27908067	.17640549
81	21.25347516	.26238858	.16140135	22.50199389	.27780239	.17553998
82	21.41405665	.26114703	.16058149	22.67668315	.27654492	.17468926
83	21.57383232	.25992569	.15977566	22.85053606	.27530766	.17385290
84	21.73281577	.25872400	.15898346	23.02356655	.27409008	.17303049
85	21.89102026	.25754141	.15820449	23.19578819	.27289163	.17222164
86	22.04845864	.25637743	.15743838	23.36721415	.27171179	.17142596
87	22.20514340	.25523153	.15668477	23.53785724	.27055008	.17064309
88	22.36108672	.25410326	.15594332	23.70772993	.26940602	.16987268
89	22.51630042	.25299214	.15521370	23.87684433	.26827915	.16911440
90	22.67079601	.25189773	.15449559	24.04521225	.26716902	.16836792
91	22.82458468	.25081961	.15378868	24.21284518	.26607522	.16763293
92	22.97767736	.24975736	.15309267	24.37975431	.26499733	.16690913
93	23.13008465	.24871059	.15240730	24.54595053	.26393495	.16619622
94	23.28181692	.24767890	.15173227	24.71144447	.26288771	.16549394
95	23.43288425	.24666194	.15106733	24.87624648	.26185523	.16480201
96	23.58329646	.24565934	.15041221	25.04036666	.26083715	.16412018
97	23.73306315	.24467075	.14976669	25.20381484	.25983314	.16344818
98	23.88219366	.24369585	.14913051	25.36660063	.25884286	.16278579
99	24.03069712	.24273431	.14850346	25.52873340	.25786599	.16213277
100	24.17858242	.24178582	.14788530	25.69022229	.25690222	.16148889

LEARNING CURVE

Percent Learning:	77.00%			78.00%		
N	CUM TOTAL	CUM AVG	UNIT	CUM TOTAL	CUM AVG	UNIT
1	1.00000000	1.00000000	1.00000000	1.00000000	1.00000000	1.00000000
2	1.77000000	.88500000	.77000000	1.78000000	.89000000	.78000000
3	2.43083351	.81027784	.66083351	2.45448763	.81816254	.67448763
4	3.02373351	.75593338	.59290000	3.06288763	.76572191	.60840000
5	3.56878763	.71375753	.54505412	3.62451909	.72490382	.56163146
6	4.07762943	.67960490	.50884180	4.15061945	.69176991	.52610035
7	4.55773769	.65110538	.48010826	4.64843820	.66406260	.49781875
8	5.01427069	.62678384	.45653300	5.12299020	.64037377	.47455200
9	5.45097161	.60566351	.43670093	5.57792377	.61976931	.45493357
10	5.87066328	.58706633	.41969167	6.01599631	.60159963	.43807254
11	6.27553965	.57050360	.40487637	6.43935517	.58539592	.42335886
12	6.66734784	.55561232	.39180819	6.84971345	.57080945	.41035828
13	7.04750727	.54211594	.38015943	7.24846518	.55757424	.39875173
14	7.41719063	.52979933	.36968336	7.63676380	.54548313	.38829863
15	7.77738066	.51849204	.36019003	8.01557728	.53437182	.37881347
16	8.12891107	.50805694	.35153041	8.38572784	.52410799	.37015056
17	8.47249675	.49838216	.34358568	8.74792138	.51458361	.36219354
18	8.80875646	.48937536	.33625971	9.10276956	.50570942	.35484818
19	9.13823022	.48095949	.32947376	9.45080678	.49741088	.34803722
20	9.46139280	.47306964	.32316259	9.79250336	.48962517	.34169658
21	9.77866443	.46565069	.31727163	10.12827595	.48229885	.33577259
22	10.09041923	.45865542	.31175480	10.45849587	.47538618	.33021991
23	10.39699213	.45204314	.30657290	10.78349578	.46884764	.32499992
24	10.69868444	.44577852	.30169230	11.10357524	.46264897	.32007946
25	10.99576843	.43983074	.29708399	11.41900514	.45676021	.31542990
26	11.28849119	.43417274	.29272276	11.73003149	.45115506	.31102635
27	11.57707780	.42878066	.28858660	12.03687855	.44581032	.30684707
28	11.86173398	.42363336	.28465619	12.33975148	.44070541	.30287293
29	12.14264845	.41871202	.28091446	12.63883855	.43582202	.29908706
30	12.41999477	.41399983	.27734632	12.93431306	.43114377	.29547451
31	12.69393307	.40948171	.27393831	13.22633499	.42665597	.29202194
32	12.96461149	.40514411	.27067842	13.51505243	.42234539	.28871744
33	13.23216736	.40097477	.26755587	13.80060275	.41820008	.28555032
34	13.49672833	.39696260	.26456097	14.08311371	.41420923	.28251096
35	13.75841332	.39309752	.26168499	14.36270438	.41036298	.27959067
36	14.01733330	.38937037	.25891998	14.63948596	.40665239	.27678158
37	14.27359206	.38577276	.25625876	14.91356250	.40306926	.27407654
38	14.52728685	.38229702	.25369479	15.18503154	.39960609	.27146903
39	14.77850895	.37893613	.25122209	15.45398465	.39625602	.26895311
40	15.02734414	.37568360	.24883519	15.72050798	.39301270	.26652333
41	15.27387322	.37253349	.24652908	15.98468268	.38987031	.26417470
42	15.51817237	.36948029	.24429915	16.24658530	.38682346	.26190262
43	15.76031353	.36651892	.24214116	16.50628817	.38386717	.25970287
44	16.00036473	.36364465	.24005120	16.76385970	.38099681	.25757153
45	16.23839037	.36085312	.23802564	17.01936471	.37820810	.25550500
46	16.47445150	.35814025	.23606113	17.27286464	.37549706	.25349994
47	16.70860608	.35550226	.23415457	17.52441786	.37285995	.25155322
48	16.94090915	.35293561	.23230307	17.77407984	.37029333	.24966198
49	17.17141309	.35043700	.23050394	18.02190335	.36779395	.24782351
50	17.40016777	.34800336	.22875467	18.26793867	.36535877	.24603532

Tables of Learning Curve Data 181

LEARNING CURVE

Percent Learning:	77.00%			78.00%		
N	CUM TOTAL	CUM AVG	UNIT	CUM TOTAL	CUM AVG	UNIT
51	17.62722070	.34563178	.22705293	18.51223373	.36298498	.24429506
52	17.85261722	.34331956	.22539653	18.75483428	.36066989	.24260055
53	18.07640064	.34106416	.22378342	18.99578403	.35841102	.24094974
54	18.29861233	.33886319	.22221169	19.23512474	.35620601	.23934071
55	18.51929186	.33671440	.22067953	19.47289640	.35405266	.23777166
56	18.73847712	.33461566	.21918526	19.70913728	.35194888	.23624088
57	18.95620442	.33256499	.21772730	19.94388408	.34989270	.23474680
58	19.17250856	.33056049	.21630414	20.17717199	.34788228	.23328791
59	19.38742293	.32860039	.21491437	20.40903479	.34591584	.23186279
60	19.60097959	.32668299	.21355667	20.63950490	.34399175	.23047012
61	19.81320936	.32480671	.21222977	20.86861353	.34210842	.22910862
62	20.02414186	.32297003	.21093250	21.09639064	.34026437	.22777711
63	20.23380558	.32117152	.20966372	21.32286510	.33845818	.22647446
64	20.44222796	.31940981	.20842238	21.54806470	.33668851	.22519960
65	20.64943543	.31768362	.20720746	21.77201622	.33495410	.22395152
66	20.85545345	.31599172	.20601802	21.99474546	.33325372	.22272925
67	21.06030658	.31433293	.20485314	22.21627735	.33158623	.22153188
68	21.26401853	.31270615	.20371195	22.43663589	.32995053	.22035855
69	21.46661218	.31111032	.20259365	22.65584432	.32834557	.21920842
70	21.66810962	.30954442	.20149744	22.87392504	.32677036	.21808072
71	21.86853220	.30800750	.20042259	23.09089974	.32522394	.21697470
72	22.06790059	.30649862	.19936838	23.30678938	.32370541	.21588963
73	22.26623474	.30501691	.19833415	23.52161423	.32221389	.21482485
74	22.46355399	.30356154	.19731925	23.73539393	.32074857	.21377970
75	22.65987704	.30213169	.19632306	23.94814750	.31930863	.21275356
76	22.85522203	.30072661	.19534499	24.15989334	.31789333	.21174585
77	23.04960652	.29934554	.19438449	24.37064932	.31650194	.21075598
78	23.24304753	.29798779	.19344101	24.58043275	.31513375	.20978343
79	23.43556158	.29665268	.19251404	24.78926042	.31378811	.20882767
80	23.62716468	.29533956	.19160310	24.99714861	.31246436	.20788820
81	23.81787237	.29404781	.19070770	25.20411317	.31116189	.20696455
82	24.00769977	.29277683	.18982739	25.41016943	.30988012	.20605627
83	24.19666151	.29152604	.18896175	25.61533234	.30861846	.20516290
84	24.38477184	.29029490	.18811035	25.81961638	.30737639	.20428405
85	24.57204465	.28908288	.18727279	26.02303567	.30615336	.20341929
86	24.75849335	.28788946	.18644869	26.22560390	.30494888	.20256824
87	24.94413104	.28671415	.18563769	26.42733443	.30376266	.20173053
88	25.12897046	.28555648	.18483942	26.62824022	.30259364	.20090580
89	25.31302401	.28441600	.18405355	26.82833392	.30144195	.20009370
90	25.49630375	.28329226	.18327974	27.02762783	.30030698	.19929390
91	25.67882143	.28218485	.18251768	27.22613392	.29918828	.19850609
92	25.86058850	.28109335	.18176707	27.42386386	.29808548	.19772995
93	26.04161612	.28001738	.18102761	27.62082905	.29699816	.19696519
94	26.22191514	.27895654	.18029902	27.81704056	.29592596	.19621151
95	26.40149617	.27791049	.17958103	28.01250922	.29486852	.19546865
96	26.58036954	.27687885	.17887337	28.20724556	.29382547	.19473634
97	26.75854532	.27586129	.17817578	28.40125988	.29279649	.19401432
98	26.93603336	.27485748	.17748804	28.59456221	.29178725	.19330234
99	27.11284324	.27386710	.17680988	28.78716237	.29077942	.19260016
100	27.28898434	.27288984	.17614110	28.97906992	.28979070	.19190755

LEARNING CURVE

Percent Learning: 79.00%   80.00%

N	CUM TOTAL	CUM AVG	UNIT	CUM TOTAL	CUM AVG	UNIT
1	1.00000000	1.00000000	1.00000000	1.00000000	1.00000000	1.00000000
2	1.79000000	.89500000	.79000000	1.80000000	.90000000	.80000000
3	2.47824455	.82608152	.68824455	2.50210370	.83403457	.70210370
4	3.10234455	.77558614	.62410000	3.14210370	.78552593	.64000000
5	3.68083669	.73616734	.57849215	3.73774105	.74754821	.59563734
6	4.22454988	.70409165	.54371319	4.29942401	.71657067	.56168296
7	4.74049429	.67721347	.51594441	4.83391353	.69055908	.53448952
8	5.23353329	.65419166	.49303900	5.34591353	.66823919	.51200000
9	5.70721385	.63413487	.47368056	5.83886314	.64876257	.49294961
10	6.16422264	.61642226	.45700880	6.31537302	.63153730	.47650987
11	6.60665606	.60060510	.44243342	6.77748416	.61613492	.46211114
12	7.03618948	.58634912	.42953342	7.22683053	.60223588	.44934637
13	7.45418846	.57339911	.41799898	7.66474605	.58959585	.43791552
14	7.86178455	.56155604	.40759608	8.09233767	.57802412	.42759162
15	8.25992861	.55066191	.39814406	8.51053685	.56736912	.41819918
16	8.64942942	.54058934	.38950081	8.92013685	.55750855	.40960000
17	9.03098213	.53123424	.38155271	9.32182029	.54834237	.40168344
18	9.40518977	.52251054	.37420764	9.71617998	.53978778	.39435969
19	9.77257974	.51434630	.36738998	10.10373493	.53177552	.38755495
20	10.13361669	.50668083	.36103695	10.48494283	.52424714	.38120790
21	10.48871262	.49946251	.35509593	10.86020990	.51715285	.37526707
22	10.83823502	.49264705	.34952240	11.22989881	.51044995	.36968891
23	11.18251344	.48619624	.34427842	11.59433505	.50410152	.36443624
24	11.52184485	.48007687	.33933140	11.95381214	.49807551	.35947710
25	11.85649801	.47425992	.33465316	12.30859599	.49234384	.35478385
26	12.18671720	.46871989	.33021919	12.65892841	.48688186	.35033242
27	12.51272526	.46343427	.32600806	13.00503015	.48166778	.34610175
28	12.83472617	.45838308	.32200091	13.34710345	.47668227	.34207330
29	13.15290725	.45354853	.31818108	13.68533413	.47190807	.33823068
30	13.46744106	.44891470	.31453381	14.01989347	.46732978	.33455935
31	13.77848698	.44446732	.31104592	14.35093980	.46293354	.33104633
32	14.08619262	.44019352	.30770564	14.67861980	.45870687	.32768000
33	14.39069501	.43608167	.30450239	15.00306974	.45463848	.32444994
34	14.69212165	.43212122	.30142664	15.32441649	.45071813	.32134675
35	14.99059144	.42830261	.29846979	15.64277841	.44693653	.31836192
36	15.28621547	.42461710	.29562403	15.95826616	.44328517	.31548775
37	15.57909776	.42105670	.29288229	16.27098339	.43975631	.31271723
38	15.86933584	.41761410	.29023808	16.58102735	.43634282	.31004396
39	16.15702136	.41428260	.28768552	16.88848946	.43303819	.30746211
40	16.44224055	.41105601	.28521919	17.19345578	.42983639	.30496632
41	16.72507468	.40792865	.28283413	17.49600746	.42673189	.30255168
42	17.00560046	.40489525	.28052578	17.79622111	.42371955	.30021366
43	17.28389039	.40195094	.27828993	18.09416921	.42079463	.29794809
44	17.56001309	.39909121	.27612270	18.38992034	.41795273	.29575113
45	17.83403357	.39631186	.27402048	18.68353953	.41518977	.29361920
46	18.10601352	.39360899	.27197995	18.97508852	.41250192	.29154899
47	18.37601155	.39097897	.26999802	19.26462596	.40988566	.28953743
48	18.64408336	.38841840	.26807181	19.55220763	.40733766	.28758168
49	18.91028199	.38592412	.26619863	19.83788669	.40485483	.28567905
50	19.17465799	.38349316	.26437600	20.12171376	.40243428	.28382708

Tables of Learning Curve Data   183

LEARNING CURVE

Percent Learning:		79.00%			80.00%		
N	CUM TOTAL	CUM AVG	UNIT	CUM TOTAL	CUM AVG	UNIT
51	19.43725956	.38112274	.26260157	20.40373719	.40007328	.28202343
52	19.69813272	.37881024	.26087316	20.68400312	.39776929	.28026593
53	19.95732146	.37655324	.25918874	20.96255568	.39551992	.27855256
54	20.21486782	.37434940	.25754637	21.23943708	.39332291	.27688140
55	20.47081208	.37219658	.25594426	21.51468773	.39117614	.27525065
56	20.72519280	.37009273	.25438072	21.78834637	.38907761	.27365864
57	20.97804694	.36803591	.25285415	22.06045013	.38702544	.27210377
58	21.22941000	.36602431	.25136306	22.33103468	.38501784	.27058454
59	21.47931602	.36405620	.24990602	22.60013424	.38305312	.26909956
60	21.72779773	.36212996	.24848171	22.86778171	.38112970	.26764748
61	21.97488659	.36024404	.24708886	23.13400876	.37924605	.26622704
62	22.22061287	.35839698	.24572628	23.39884582	.37740074	.26483706
63	22.46500571	.35658739	.24439283	23.66232222	.37559242	.26347640
64	22.70809316	.35481396	.24308746	23.92446622	.37381978	.26214400
65	22.94990229	.35307542	.24180913	24.18530506	.37208162	.26083884
66	23.19045917	.35137059	.24055689	24.44486501	.37037674	.25955995
67	23.42978899	.34969834	.23932981	24.70317144	.36870405	.25830643
68	23.66791603	.34805759	.23812704	24.96024884	.36706248	.25707740
69	23.90486378	.34644730	.23694775	25.21612087	.36545103	.25587203
70	24.14065491	.34486650	.23579113	25.47081041	.36386872	.25468954
71	24.37531137	.34331424	.23465645	25.72433957	.36231464	.25352917
72	24.60885435	.34178964	.23354299	25.97672977	.36078791	.25239020
73	24.84130441	.34029184	.23245006	26.22800173	.35928769	.25127196
74	25.07268142	.33882002	.23137701	26.47817551	.35781318	.25017378
75	25.30300463	.33737340	.23032321	26.72727056	.35636361	.24909505
76	25.53229272	.33595122	.22928808	26.97530573	.35493823	.24803517
77	25.76056376	.33455278	.22827105	27.22229929	.35353635	.24699356
78	25.98783532	.33317738	.22727156	27.46826898	.35215729	.24596969
79	26.21412442	.33182436	.22628910	27.71323200	.35080041	.24496302
80	26.43944758	.33049309	.22532316	27.95720506	.34946506	.24397306
81	26.66382085	.32918297	.22437327	28.20020437	.34815067	.24299932
82	26.88725981	.32789341	.22343896	28.44224571	.34685666	.24204134
83	27.10977962	.32662385	.22251980	28.68334440	.34558246	.24109869
84	27.33139498	.32537375	.22161537	28.92351533	.34432756	.24017093
85	27.55212023	.32414259	.22072524	29.16277298	.34309145	.23925765
86	27.77196927	.32292988	.21984905	29.40113146	.34187362	.23835848
87	27.99095567	.32173512	.21898639	29.63860447	.34067361	.23747301
88	28.20909260	.32055787	.21813693	29.87520538	.33949097	.23660090
89	28.42639290	.31939767	.21730030	30.11094717	.33832525	.23574180
90	28.64286908	.31825410	.21647618	30.34584253	.33717603	.23489536
91	28.85853332	.31712674	.21566424	30.57990379	.33604290	.23406126
92	29.07339748	.31601519	.21486416	30.81314298	.33492547	.23323919
93	29.28747314	.31491907	.21407566	31.04557183	.33382335	.23242885
94	29.50077158	.31383800	.21329844	31.27720178	.33273619	.23162995
95	29.71330380	.31277162	.21253222	31.50804398	.33166362	.23084220
96	29.92508053	.31171959	.21177673	31.73810932	.33060531	.23006534
97	30.13611224	.31068157	.21103171	31.96740843	.32956091	.22929911
98	30.34640916	.30965724	.21029692	32.19595161	.32853012	.22854324
99	30.55598127	.30864628	.20957211	32.42374917	.32751262	.22779751
100	30.76483831	.30764838	.20885704	32.65081083	.32650811	.22706166

LEARNING CURVE

Percent Learning:	81.00%			82.00%		
N	CUM TOTAL	CUM AVG	UNIT	CUM TOTAL	CUM AVG	UNIT
1	1.00000000	1.00000000	1.00000000	1.00000000	1.00000000	1.00000000
2	1.81000000	.90500000	.81000000	1.82000000	.91000000	.82000000
3	2.52606457	.84202152	.71606457	2.55012663	.85004221	.73012663
4	3.18216457	.79554114	.65610000	3.22252663	.80563166	.67240000
5	3.79523278	.75904656	.61306821	3.85331251	.77066250	.63078588
6	4.37524508	.72920751	.58001230	4.45201634	.74200272	.59870384
7	4.92870346	.70410049	.55345838	5.02487159	.71783880	.57285525
8	5.46014446	.68251806	.53144100	5.57623959	.69702995	.55136800
9	5.97289293	.66365477	.51274847	6.10932448	.67881383	.53308489
10	6.46947818	.64694782	.49658525	6.62656890	.66265689	.51724442
11	6.95188138	.63198922	.48240320	7.12988975	.64817180	.50332084
12	7.42169134	.61847428	.46980997	7.62082689	.63506891	.49093715
13	7.88020717	.60616978	.45851582	8.10064138	.62312626	.47981449
14	8.32850846	.59489346	.44830129	8.57038268	.61217019	.46974130
15	8.76750488	.58450033	.43899642	9.03093625	.60206242	.46055357
16	9.19797209	.57487326	.43046721	9.48305801	.59269113	.45212176
17	9.62057834	.56591637	.42260625	9.92739996	.58396470	.44434196
18	10.03590460	.55755026	.41532626	10.36452958	.57580720	.43712961
19	10.44446003	.54970842	.40855544	10.79494467	.56815498	.43041509
20	10.84669409	.54233470	.40223405	11.21908510	.56095425	.42414043
21	11.24300602	.53538124	.39631194	11.63734197	.55415914	.41825687
22	11.63375262	.52880694	.39074659	12.05006506	.54773023	.41272309
23	12.01925433	.52257628	.38550171	12.45756882	.54163343	.40750376
24	12.39980040	.51665835	.38054607	12.86013728	.53583905	.40256846
25	12.77565303	.51102612	.37585263	13.25802810	.53032112	.39789083
26	13.14705084	.50565580	.37139782	13.65147598	.52505677	.39344788
27	13.51421186	.50052637	.36716101	14.04069546	.52002576	.38921948
28	13.87733590	.49561914	.36312404	14.42588332	.51521012	.38518787
29	14.23660673	.49091747	.35927083	14.80722066	.51059382	.38133734
30	14.59219384	.48640646	.35558710	15.18487459	.50616249	.37765393
31	14.94425394	.48207271	.35206011	15.55899974	.50190322	.37412515
32	15.29293238	.47790414	.34867844	15.92973958	.49780436	.37073984
33	15.63836423	.47388983	.34543184	16.29722753	.49385538	.36748795
34	15.98067529	.47001986	.34231106	16.66158794	.49004670	.36436041
35	16.31998302	.46628523	.33930774	17.02293694	.48636963	.36134900
36	16.65639729	.46267770	.33641427	17.38138322	.48281620	.35844628
37	16.99002106	.45918976	.33362376	17.73702869	.47937915	.35564547
38	17.32095096	.45581450	.33092990	18.08996907	.47605182	.35294036
39	17.64927790	.45254559	.32832694	18.44029440	.47282806	.35032533
40	17.97508748	.44937719	.32580958	18.78808955	.46970224	.34779515
41	18.29846046	.44630391	.32337298	19.13343460	.46666914	.34534505
42	18.61947313	.44332079	.32101267	19.47640523	.46372393	.34297063
43	18.93819765	.44042320	.31872453	19.81707308	.46086216	.34066785
44	19.25470239	.43760687	.31650474	20.15550601	.45807968	.33843294
45	19.56905218	.43486783	.31434978	20.49176844	.45537263	.33626242
46	19.88130857	.43220236	.31225639	20.82592152	.45273742	.33415308
47	20.19153007	.42960702	.31022151	21.15802344	.45017071	.33210192
48	20.49977239	.42707859	.30824232	21.48812958	.44766937	.33010614
49	20.80608857	.42461405	.30631618	21.81629271	.44523046	.32816313
50	21.11052920	.42221058	.30444063	22.14256319	.44285126	.32627048

LEARNING CURVE

Percent Learning: 81.00%   82.00%

N	CUM TOTAL	CUM AVG	UNIT	CUM TOTAL	CUM AVG	UNIT
51	21.41314256	.41986554	.30261336	22.46698908	.44052920	.32442590
52	21.71397479	.41757644	.30083223	22.78961634	.43826185	.32262726
53	22.01307001	.41534094	.29909522	23.11048892	.43604696	.32087258
54	22.31047043	.41315686	.29740042	23.42964889	.43388239	.31915997
55	22.60621650	.41102212	.29574607	23.74713657	.43176612	.31748768
56	22.90034697	.40893477	.29413047	24.06299062	.42969626	.31585405
57	23.19289904	.40689297	.29255207	24.37724814	.42767102	.31425752
58	23.48390842	.40489497	.29100937	24.68994476	.42568870	.31269662
59	23.77340940	.40293914	.28950098	25.00111472	.42374771	.31116996
60	24.06143495	.40102392	.28802555	25.31079094	.42184652	.30967622
61	24.34801680	.39914782	.28658185	25.61900510	.41998369	.30821417
62	24.63318549	.39730944	.28516869	25.92578773	.41815787	.30678262
63	24.91697042	.39550747	.28378494	26.23116820	.41636775	.30538048
64	25.19939996	.39374062	.28242954	26.53517487	.41461211	.30400667
65	25.48050144	.39200771	.28110147	26.83783508	.41288977	.30266020
66	25.76030123	.39030759	.27979979	27.13917520	.41119962	.30134012
67	26.03882480	.38863918	.27852358	27.43922071	.40954061	.30004552
68	26.31609676	.38700142	.27727196	27.73799625	.40791171	.29877553
69	26.59214088	.38539335	.27604412	28.03552559	.40631197	.29752935
70	26.86698015	.38381400	.27483927	28.33183177	.40474045	.29630618
71	27.14063680	.38226249	.27365665	28.62693706	.40319630	.29510528
72	27.41313236	.38073795	.27249556	28.92086301	.40167865	.29392595
73	27.68448767	.37923956	.27135531	29.21363051	.40018672	.29276750
74	27.95472291	.37776653	.27023525	29.50525980	.39871973	.29162929
75	28.22385766	.37631810	.26913475	29.79577048	.39727694	.29051069
76	28.49191089	.37489356	.26805322	30.08518519	.39585765	.28941111
77	28.75890098	.37349222	.26699009	30.37351158	.39446119	.28832999
78	29.02484580	.37211341	.26594482	30.66077835	.39308690	.28726677
79	29.28976267	.37075649	.26491688	30.94699930	.39173417	.28622095
80	29.55366843	.36942086	.26390576	31.23219133	.39040239	.28519202
81	29.81657943	.36810592	.26291099	31.51637083	.38909100	.28417950
82	30.07851154	.36681112	.26193211	31.79955377	.38779944	.28318294
83	30.33948022	.36553591	.26096868	32.08175565	.38652718	.28220189
84	30.59950048	.36427977	.26002026	32.36299157	.38527371	.28123592
85	30.85858694	.36304220	.25908646	32.64327620	.38403854	.28028463
86	31.11675380	.36182272	.25816687	32.92262384	.38282121	.27934763
87	31.37401492	.36062086	.25726112	33.20104839	.38162125	.27842455
88	31.63038376	.35943618	.25636884	33.47856339	.38043822	.27751501
89	31.88587345	.35826824	.25548969	33.75518206	.37927171	.27661867
90	32.14049677	.35711663	.25462333	34.03091725	.37812130	.27573519
91	32.39426620	.35598095	.25376943	34.30578149	.37698661	.27486425
92	32.64719387	.35486080	.25292767	34.57978702	.37586725	.27400553
93	32.89929164	.35375582	.25209777	34.85294576	.37476286	.27315874
94	33.15057107	.35266565	.25127942	35.12526933	.37367308	.27232357
95	33.40104341	.35158993	.25047235	35.39676909	.37259757	.27149976
96	33.65071969	.35052833	.24967628	35.66745612	.37153600	.27068703
97	33.89961064	.34948052	.24889095	35.93734124	.37048805	.26988512
98	34.14772674	.34844619	.24811610	36.20643501	.36945342	.26909377
99	34.39507825	.34742503	.24735150	36.47474775	.36843180	.26831274
100	34.64167515	.34641675	.24659691	36.74228954	.36742290	.26754179

LEARNING CURVE

Percent Learning:	83.00%			84.00%		
N	CUM TOTAL	CUM AVG	UNIT	CUM TOTAL	CUM AVG	UNIT
1	1.00000000	1.00000000	1.00000000	1.00000000	1.00000000	1.00000000
2	1.83000000	.91500000	.83000000	1.84000000	.92000000	.84000000
3	2.57428936	.85809645	.74428936	2.59855226	.86618409	.75855226
4	3.26318936	.81579734	.68890000	3.30415226	.82603807	.70560000
5	3.91198086	.78239617	.64879150	3.97123845	.79424769	.66708619
6	4.52974103	.75495684	.61776017	4.60842235	.76807039	.63718390
7	5.12242541	.73177506	.59268438	5.22137239	.74591034	.61295004
8	5.69421241	.71177655	.57178700	5.81407639	.72675955	.59270400
9	6.24817906	.69424212	.55396665	6.38947793	.70994199	.57540153
10	6.78667600	.67866760	.53849694	6.94983032	.69498303	.56035240
11	7.31155138	.66468649	.52487537	7.49690847	.68153713	.54707815
12	7.82429232	.65202436	.51274094	8.03214295	.66934525	.53523448
13	8.32611855	.64047066	.50182623	8.55670886	.65820837	.52456591
14	8.81804659	.62986047	.49192804	9.07158689	.64797049	.51487804
15	9.30093519	.62006235	.48288861	9.57760663	.63850711	.50601973
16	9.77551840	.61096990	.47458321	10.07547799	.62971737	.49787136
17	10.24243005	.60249589	.46691165	10.56581468	.62151851	.49033669
18	10.70222237	.59456791	.45979232	11.04915197	.61384178	.48333729
19	11.15538032	.58712528	.45315795	11.52596037	.60662949	.47680840
20	11.60233279	.58011664	.44695246	11.99665638	.59983282	.47069601
21	12.04346147	.57349817	.44112868	12.46161103	.59341005	.46495464
22	12.47910802	.56723218	.43564656	12.92115667	.58732530	.45954564
23	12.90957985	.56128608	.43047183	13.37559260	.58154750	.45443593
24	13.33515483	.55563145	.42557498	13.82518956	.57604956	.44959696
25	13.75608524	.55024341	.42093041	14.27019354	.57080774	.44500398
26	14.17260101	.54510024	.41651577	14.71082890	.56580111	.44063536
27	14.58491250	.54018194	.41231149	15.14730103	.56101115	.43647214
28	14.99321277	.53547188	.40830027	15.57979859	.55642138	.43249755
29	15.39767960	.53095447	.40446683	16.00849536	.55201708	.42869677
30	15.79847714	.52661590	.40079754	16.43355193	.54778506	.42505658
31	16.19575740	.52244379	.39728026	16.85511710	.54371345	.42156517
32	16.58966147	.51842692	.39390406	17.27332904	.53979153	.41821194
33	16.98032062	.51455517	.39065915	17.68831641	.53600959	.41498737
34	17.36785729	.51081933	.38753667	18.10019923	.53235880	.41188282
35	17.75238588	.50721103	.38452859	18.50908974	.52883114	.40889051
36	18.13401350	.50372260	.38162763	18.91509306	.52541925	.40600332
37	18.51284065	.50034704	.37882715	19.31830787	.52211643	.40321481
38	18.88896175	.49707794	.37612110	19.71882692	.51891650	.40051905
39	19.26246568	.49390938	.37350392	20.11673758	.51581378	.39791066
40	19.63343622	.49083591	.37097054	20.51212223	.51280306	.39538465
41	20.00195249	.48785250	.36851627	20.90505870	.50987948	.39293647
42	20.36808930	.48495451	.36613681	21.29562060	.50703859	.39056190
43	20.73191745	.48213762	.36382816	21.68387766	.50427622	.38825706
44	21.09350410	.47939782	.36158664	22.06989600	.50158855	.38601834
45	21.45291295	.47673140	.35940885	22.45373841	.49897196	.38384241
46	21.81020457	.47413488	.35729162	22.83546459	.49642314	.38172618
47	22.16543656	.47160503	.35523199	23.21513134	.49393896	.37966675
48	22.51866380	.46913883	.35322723	23.59279279	.49151652	.37766145
49	22.86993858	.46673344	.35127478	23.96850055	.48915307	.37570776
50	23.21931081	.46438622	.34937224	24.34230389	.48684608	.37380334

LEARNING CURVE

Percent Learning:	83.00%			84.00%		
N	CUM TOTAL	CUM AVG	UNIT	CUM TOTAL	CUM AVG	UNIT
51	23.56682819	.46209467	.34751737	24.71424990	.48459314	.37194601
52	23.91253628	.45985647	.34570809	25.08438360	.48239199	.37013371
53	24.25647870	.45766941	.34394243	25.45274811	.48024053	.36836450
54	24.59869724	.45553143	.34221853	25.81938470	.47813675	.36663659
55	24.93923191	.45344058	.34053468	26.18433298	.47607878	.36494827
56	25.27812114	.45139502	.33888923	26.54763092	.47406484	.36329794
57	25.61540178	.44939301	.33728064	26.90931501	.47209325	.36168409
58	25.95110925	.44743292	.33570747	27.26942029	.47016242	.36010529
59	26.28527759	.44551318	.33416834	27.62798048	.46827086	.35856019
60	26.61793955	.44363233	.33266196	27.98502801	.46641713	.35704752
61	26.94912666	.44178896	.33118711	28.34059410	.46459990	.35556609
62	27.27886927	.43998176	.32974261	28.69470884	.46281788	.35411474
63	27.60719665	.43820947	.32832738	29.04740123	.46106986	.35269240
64	27.93413703	.43647089	.32694037	29.39869926	.45935468	.35129803
65	28.25971762	.43476489	.32558059	29.74862994	.45767123	.34993067
66	28.58396472	.43309037	.32424710	30.09721932	.45601847	.34858939
67	28.90690371	.43144632	.32293900	30.44449263	.45439541	.34727330
68	29.22855915	.42983175	.32165544	30.79047420	.45280109	.34598157
69	29.54895475	.42824572	.32039560	31.13518760	.45123460	.34471340
70	29.86811348	.42668734	.31915873	31.47865562	.44969508	.34346803
71	30.18605755	.42515574	.31794407	31.82090034	.44818169	.34224472
72	30.50280848	.42365012	.31675093	32.16194314	.44669365	.34104279
73	30.81838711	.42216969	.31557863	32.50180471	.44523020	.33986157
74	31.13281365	.42071370	.31442653	32.84050515	.44379061	.33870044
75	31.44610767	.41928144	.31329402	33.17806392	.44237419	.33755877
76	31.75828818	.41787221	.31218051	33.51449993	.44098026	.33643601
77	32.06937362	.41648537	.31108544	33.84983150	.43960820	.33533158
78	32.37938188	.41512028	.31000826	34.18407646	.43825739	.33424495
79	32.68833034	.41377633	.30894846	34.51725208	.43692724	.33317563
80	32.99623589	.41245295	.30790555	34.84937519	.43561719	.33212311
81	33.30311494	.41114957	.30687905	35.18046211	.43432669	.33108693
82	33.60898345	.40986565	.30586851	35.51052875	.43305523	.33006663
83	33.91385693	.40860069	.30487348	35.83959054	.43180230	.32906180
84	34.21775048	.40735417	.30389355	36.16766254	.43056741	.32807200
85	34.52067878	.40612563	.30292831	36.49475937	.42935011	.32709684
86	34.82265615	.40491461	.30197737	36.82089530	.42814995	.32613593
87	35.12369651	.40372065	.30104036	37.14608421	.42696649	.32518890
88	35.42381342	.40254333	.30011691	37.47033961	.42579931	.32425541
89	35.72302011	.40138225	.29920669	37.79367470	.42464803	.32333509
90	36.02132946	.40023699	.29830935	38.11610233	.42351225	.32242763
91	36.31875403	.39910719	.29742457	38.43763503	.42239159	.32153270
92	36.61530607	.39799246	.29655204	38.75828502	.42128571	.32064999
93	36.91099754	.39689245	.29569147	39.07806423	.42019424	.31977921
94	37.20584010	.39580681	.29484256	39.39698430	.41911685	.31892007
95	37.49984512	.39473521	.29400502	39.71505660	.41805323	.31807230
96	37.79302373	.39367733	.29317860	40.03229221	.41700304	.31723561
97	38.08538676	.39263285	.29236304	40.34870198	.41596600	.31640977
98	38.37694483	.39160148	.29155807	40.66429650	.41494180	.31559452
99	38.66770828	.39058291	.29076345	40.97908611	.41393016	.31478961
100	38.95768724	.38957687	.28997896	41.29308091	.41293081	.31399481

LEARNING CURVE

Percent Learning:		85.00%				86.00%	
N	CUM TOTAL	CUM AVG	UNIT	CUM TOTAL	CUM AVG	UNIT
---	---	---	---	---	---	---
1	1.00000000	1.00000000	1.00000000	1.00000000	1.00000000	1.00000000
2	1.85000000	.92500000	.85000000	1.86000000	.93000000	.86000000
3	2.62291484	.87430495	.77291484	2.64737660	.88245887	.78737660
4	3.34541484	.83635371	.72250000	3.38697660	.84674415	.73960000
5	4.03108590	.80621718	.68567106	4.09152383	.81830477	.70454723
6	4.68806351	.78134392	.65697761	4.76866770	.79477795	.67714387
7	5.32171998	.76024571	.63365647	5.42347559	.77478223	.65480789
8	5.93584498	.74198062	.61412500	6.05953159	.75744145	.63605600
9	6.53324232	.72591581	.59739734	6.67949350	.74216594	.61996190
10	7.11606273	.71160627	.58282040	7.28540412	.72854041	.60591062
11	7.68600333	.69872758	.56994061	7.87887835	.71626167	.59347424
12	8.24443430	.68703619	.55843097	8.46122208	.70510184	.58234373
13	8.79248280	.67634483	.54804849	9.03351117	.69488547	.57228909
14	9.33109080	.66650649	.53860800	9.59664596	.68547471	.56313479
15	9.86105613	.65740374	.52996534	10.15138996	.67675933	.55474400
16	10.38306238	.64894140	.52200625	10.69839812	.66864988	.54700816
17	10.89770113	.64104124	.51463875	11.23823787	.66107282	.53983974
18	11.40548887	.63363827	.50778774	11.77140510	.65396695	.53316724
19	11.90688008	.62667790	.50139121	12.29833661	.64728087	.52693151
20	12.40227742	.62011387	.49539734	12.81941975	.64097099	.52108313
21	12.89203991	.61390666	.48976249	13.33500015	.63500001	.51558041
22	13.37648942	.60802225	.48444952	13.84538800	.62933582	.51038784
23	13.85591604	.60243113	.47942661	14.35086300	.62395057	.50547501
24	14.33058236	.59710760	.47466632	14.85167861	.61881994	.50081561
25	14.80072717	.59202909	.47014480	15.34806542	.61392262	.49638680
26	15.26656839	.58717571	.46584122	15.84023403	.60923977	.49216862
27	15.72830566	.58252984	.46173727	16.32837753	.60475472	.48814349
28	16.18612246	.57807580	.45781680	16.81267344	.60045262	.48429592
29	16.64018793	.57379958	.45406548	17.29328557	.59632019	.48061212
30	17.09065847	.56968862	.45047054	17.77036541	.59234551	.47707984
31	17.53767903	.56573158	.44702056	18.24405350	.58851785	.47368809
32	17.98138435	.56191826	.44370531	18.71448052	.58482752	.47042702
33	18.42189990	.55823939	.44051555	19.18176824	.58126570	.46728772
34	18.85934283	.55468655	.43744294	19.64603042	.57782442	.46426218
35	19.29382274	.55125208	.43447990	20.10737351	.57449639	.46134309
36	19.72544232	.54792895	.43161958	20.56589733	.57127493	.45852382
37	20.15429802	.54471076	.42885570	21.02169567	.56815394	.45579834
38	20.58048054	.54159159	.42618253	21.47485677	.56512781	.45316110
39	21.00407535	.53856603	.42359481	21.92546380	.56219138	.45060704
40	21.42516310	.53562908	.42108774	22.37359530	.55933988	.44813149
41	21.84381997	.53277610	.41865687	22.81932548	.55656891	.44573019
42	22.26011808	.53000281	.41629811	23.26272464	.55387440	.44339915
43	22.67412577	.52730525	.41400769	23.70385937	.55125254	.44113473
44	23.08590786	.52467972	.41178209	24.14279292	.54869984	.43893355
45	23.49552593	.52212280	.40961807	24.57958536	.54621301	.43679244
46	23.90303855	.51963127	.40751262	25.01429386	.54378900	.43470851
47	24.30850147	.51720216	.40546293	25.44697287	.54142495	.43267901
48	24.71196785	.51483266	.40346638	25.87767430	.53911821	.43070142
49	25.11348837	.51252017	.40152052	26.30644767	.53686628	.42877337
50	25.51311146	.51026223	.39962308	26.73334032	.53466681	.42689265

LEARNING CURVE

Percent Learning:		85.00%				86.00%	
N	CUM TOTAL	CUM AVG	UNIT	CUM TOTAL	CUM AVG	UNIT
51	25.91088338	.50805654	.39777192	27.15839750	.53251760	.42505718
52	26.30684842	.50590093	.39596504	27.58166251	.53041659	.42326501
53	26.70104896	.50379338	.39420055	28.00317684	.52836183	.42151433
54	27.09352564	.50173196	.39247668	28.42298024	.52635149	.41980340
55	27.48431742	.49971486	.39079178	28.84111088	.52438383	.41813063
56	27.87346170	.49774039	.38914428	29.25760536	.52245724	.41649449
57	28.26099441	.49580692	.38753270	29.67249890	.52057016	.41489354
58	28.64695006	.49391293	.38595565	30.08582533	.51872113	.41332643
59	29.03136188	.49205698	.38441182	30.49761720	.51690877	.41179187
60	29.41426184	.49023770	.38289996	30.90790586	.51513176	.41028866
61	29.79568072	.48845378	.38141888	31.31672152	.51338888	.40881566
62	30.17564820	.48670400	.37996748	31.72409328	.51167892	.40737176
63	30.55419289	.48498719	.37854469	32.13004922	.51000078	.40595595
64	30.93134241	.48330223	.37714952	32.53461646	.50835338	.40456724
65	31.30712340	.48164805	.37578099	32.93782116	.50673571	.40320470
66	31.68156162	.48002366	.37443822	33.33968860	.50514680	.40186744
67	32.05468194	.47842809	.37312033	33.74024324	.50358572	.40055464
68	32.42650844	.47686042	.37182650	34.13950871	.50205160	.39926547
69	32.79706438	.47531977	.37055594	34.53750790	.50054359	.39799919
70	33.16637230	.47380532	.36930792	34.93426295	.49906090	.39675506
71	33.53445401	.47231625	.36808171	35.32979533	.49760275	.39553238
72	33.90133066	.47085181	.36687664	35.72412582	.49616841	.39433049
73	34.26702272	.46941127	.36569206	36.11727457	.49475719	.39314876
74	34.63155006	.46799392	.36452734	36.50926114	.49336839	.39198657
75	34.99493196	.46659909	.36338189	36.90010450	.49200139	.39084335
76	35.35718710	.46522615	.36225515	37.28982304	.49065557	.38971854
77	35.71833366	.46387446	.36114655	37.67843465	.48933032	.38861161
78	36.07838925	.46254345	.36005559	38.06595670	.48802509	.38752205
79	36.43737101	.46123254	.35898176	38.45240607	.48673932	.38644936
80	36.79529559	.45994119	.35792458	38.83779916	.48547249	.38539309
81	37.15217917	.45866888	.35688359	39.22215192	.48422410	.38435276
82	37.50803751	.45741509	.35585834	39.60547988	.48299366	.38332796
83	37.86288593	.45617935	.35484841	39.98779814	.48178070	.38231826
84	38.21673933	.45496118	.35385340	40.36912141	.48058478	.38132327
85	38.56961222	.45376014	.35287290	40.74946401	.47940546	.38034260
86	38.92151876	.45257580	.35190653	41.12883988	.47824232	.37937587
87	39.27247270	.45140773	.35095394	41.50726261	.47709497	.37842274
88	39.62248747	.45025554	.35001477	41.88474546	.47596302	.37748285
89	39.97157616	.44911883	.34908869	42.26130134	.47484608	.37655588
90	40.31975152	.44799724	.34817536	42.63694284	.47374381	.37564150
91	40.66702600	.44689040	.34727447	43.01168226	.47265585	.37473941
92	41.01341173	.44579795	.34638573	43.38553157	.47158186	.37384931
93	41.35892055	.44471958	.34550883	43.75850249	.47052153	.37297092
94	41.70356404	.44365494	.34464349	44.13060644	.46947454	.37210395
95	42.04735348	.44260372	.34378944	44.50185457	.46844057	.37124814
96	42.39029990	.44156562	.34294642	44.87225780	.46741935	.37040322
97	42.73241407	.44054035	.34211417	45.24182676	.46641059	.36956896
98	43.07370651	.43952762	.34129244	45.61057186	.46541400	.36874510
99	43.41418752	.43852715	.34048100	45.97850328	.46442933	.36793142
100	43.75386714	.43753867	.33967962	46.34563096	.46345631	.36712768

LEARNING CURVE

Percent Learning:	87.00%			88.00%		
N	CUM TOTAL	CUM AVG	UNIT	CUM TOTAL	CUM AVG	UNIT
1	1.00000000	1.00000000	1.00000000	1.00000000	1.00000000	1.00000000
2	1.87000000	.93500000	.87000000	1.88000000	.94000000	.88000000
3	2.67193706	.89064569	.80193706	2.69659575	.89886525	.81659575
4	3.42883706	.85720926	.75690000	3.47099575	.86774894	.77440000
5	4.15255286	.83051057	.72371580	4.21417360	.84283472	.74317785
6	4.85023810	.80837302	.69768524	4.93277787	.82212964	.71860426
7	5.52664663	.78952095	.67640853	5.63124047	.80446292	.69846260
8	6.18514963	.77314370	.65850300	6.31271247	.78908906	.68147200
9	6.82825268	.75869474	.64310305	6.97954109	.77550457	.66682862
10	7.45788542	.74578854	.62963274	7.63353760	.76335376	.65399651
11	8.07557602	.73414327	.61769060	8.27613890	.75237626	.64260130
12	8.68256218	.72354685	.60698616	8.90851065	.74237589	.63237175
13	9.27986510	.71383578	.59730292	9.53161599	.73320123	.62310534
14	9.86834052	.70488147	.58847542	10.14626308	.72473308	.61464709
15	10.44871504	.69658100	.58037452	10.75313896	.71687593	.60687588
16	11.02161265	.68885079	.57289761	11.35283432	.70955214	.59969536
17	11.58757455	.68162203	.56596190	11.94586203	.70269777	.59302771
18	12.14707420	.67483746	.55949965	12.53267122	.69625951	.58680919
19	12.70052901	.66844890	.55345481	13.11365822	.69019254	.58098700
20	13.24830950	.66241547	.54778049	13.68917514	.68445876	.57551693
21	13.79074657	.65670222	.54243707	14.25953674	.67902556	.57036159
22	14.32813739	.65127897	.53739082	14.82502588	.67386421	.56548915
23	14.86075018	.64611957	.53261279	15.38589810	.66895209	.56087222
24	15.38882814	.64120117	.52807796	15.94238524	.66426605	.55648714
25	15.91259269	.63650371	.52376456	16.49469856	.65978794	.55231332
26	16.43224624	.63200947	.51965354	17.04303126	.65550120	.54833270
27	16.94797440	.62770276	.51572817	17.58756068	.65139114	.54452942
28	17.45994802	.62356957	.51197362	18.12845012	.64744465	.54088944
29	17.96832477	.61959741	.50837675	18.66585038	.64365001	.53740027
30	18.47325060	.61577502	.50492583	19.19990115	.63999671	.53405077
31	18.97486097	.61209229	.50161037	19.73073213	.63647523	.53083098
32	19.47328189	.60854006	.49842092	20.25846405	.63307700	.52773192
33	19.96863087	.60511003	.49534898	20.78320954	.62979423	.52474549
34	20.46101772	.60179464	.49238685	21.30507392	.62661982	.52186438
35	20.95054526	.59858701	.48952754	21.82415586	.62354731	.51908193
36	21.43730995	.59548083	.48676470	22.34054794	.62057078	.51639209
37	21.92140247	.59247034	.48409251	22.85433726	.61768479	.51378932
38	22.40290816	.58955021	.48150569	23.36560582	.61488436	.51126856
39	22.88190750	.58671558	.47899935	23.87443100	.61216490	.50882518
40	23.35847653	.58396191	.47656902	24.38088589	.60952215	.50645490
41	23.83268712	.58128505	.47421059	24.88503968	.60695219	.50415378
42	24.30460737	.57868113	.47192025	25.38695788	.60445138	.50191820
43	24.77430185	.57614655	.46969448	25.88670267	.60201634	.49974480
44	25.24183186	.57367800	.46753001	26.38433312	.59964393	.49763045
45	25.70725569	.57127235	.46542383	26.87990538	.59733123	.49557226
46	26.17062882	.56892671	.46337313	27.37347294	.59507550	.49356755
47	26.63200409	.56663838	.46137527	27.86508674	.59287419	.49161380
48	27.09143192	.56440483	.45942782	28.35479543	.59072490	.48970868
49	27.54896042	.56222368	.45752850	28.84264543	.58862542	.48785000
50	28.00463558	.56009271	.45567517	29.32868115	.58657362	.48603572

LEARNING CURVE

Percent Learning:	87.00%			88.00%		
N	CUM TOTAL	CUM AVG	UNIT	CUM TOTAL	CUM AVG	UNIT
51	28.45850140	.55800983	.45386582	29.81294505	.58456755	.48426391
52	28.91059999	.55597308	.45209858	30.29547783	.58260534	.48253278
53	29.36097168	.55398060	.45037170	30.77631847	.58068525	.48084064
54	29.80965519	.55203065	.44868350	31.25550436	.57880564	.47918589
55	30.25668763	.55012159	.44703244	31.73307141	.57696493	.47756705
56	30.70210468	.54825187	.44541705	32.20905412	.57516168	.47598271
57	31.14594060	.54642001	.44383593	32.68348564	.57339448	.47443152
58	31.58822837	.54462463	.44228777	33.15639787	.57166203	.47291223
59	32.02899972	.54286440	.44077134	33.62782153	.56996308	.47142366
60	32.46828519	.54113809	.43928547	34.09778621	.56829644	.46996468
61	32.90611424	.53944450	.43782905	34.56632042	.56666099	.46853421
62	33.34251526	.53778250	.43640102	35.03345168	.56505567	.46713126
63	33.77751565	.53615104	.43500039	35.49920654	.56347947	.46575486
64	34.21114185	.53454909	.43362620	35.96361063	.56193142	.46440409
65	34.64341941	.53297568	.43227756	36.42668872	.56041060	.46307809
66	35.07437302	.53142989	.43095361	36.88846475	.55891613	.46177603
67	35.50402656	.52991084	.42965354	37.34896189	.55744719	.46049714
68	35.93240312	.52841769	.42837656	37.80820255	.55600298	.45924066
69	36.35952505	.52694964	.42712194	38.26620842	.55458273	.45800587
70	36.78541401	.52550591	.42588896	38.72300052	.55318572	.45679210
71	37.21009097	.52408579	.42467696	39.17859922	.55181126	.45559870
72	37.63357626	.52268856	.42348529	39.63302425	.55045867	.45442504
73	38.05588958	.52131356	.42231332	40.08629478	.54912733	.45327053
74	38.47705007	.51996014	.42116049	40.53842938	.54781661	.45213460
75	38.89707628	.51862768	.42002621	40.98944609	.54652595	.45101671
76	39.31598623	.51731561	.41890995	41.43936242	.54525477	.44991633
77	39.73379742	.51602334	.41781119	41.88819539	.54400254	.44883298
78	40.15052685	.51475034	.41672943	42.33596155	.54276874	.44776616
79	40.56619106	.51349609	.41566421	42.78267696	.54155287	.44671541
80	40.98080611	.51226008	.41461505	43.22835727	.54035447	.44568031
81	41.39438764	.51104182	.41358153	43.67301769	.53917306	.44466041
82	41.80695085	.50984086	.41256322	44.11667301	.53800821	.44365533
83	42.21851056	.50865675	.41155971	44.55933767	.53685949	.44266465
84	42.62908118	.50748906	.41057062	45.00102569	.53572650	.44168802
85	43.03867674	.50633737	.40959557	45.44175074	.53460883	.44072506
86	43.44731094	.50520129	.40863419	45.88152616	.53350612	.43977542
87	43.85499709	.50408043	.40768615	46.32036494	.53241799	.43883877
88	44.26174820	.50297441	.40675111	46.75827973	.53134409	.43791479
89	44.66757694	.50188289	.40582874	47.19528290	.53028408	.43700317
90	45.07249568	.50080551	.40491874	47.63138649	.52923763	.43610359
91	45.47651647	.49974194	.40402079	48.06660227	.52820442	.43521578
92	45.87965109	.49869186	.40313462	48.50094171	.52718415	.43433945
93	46.28191104	.49765496	.40225994	48.93441604	.52617652	.43347432
94	46.68330752	.49663093	.40139649	49.36703618	.52518124	.43262015
95	47.08385152	.49561949	.40054399	49.79881285	.52419803	.43177667
96	47.48355372	.49462035	.39970221	50.22975650	.52322663	.43094364
97	47.88242461	.49363324	.39887089	50.65987733	.52226678	.43012083
98	48.28047441	.49265790	.39804980	51.08918533	.52131822	.42930800
99	48.67771311	.49169407	.39723870	51.51769027	.52038071	.42850494
100	49.07415051	.49074151	.39643739	51.94540170	.51945402	.42771143

LEARNING CURVE

Percent Learning:		89.00%			90.00%	
N	CUM TOTAL	CUM AVG	UNIT	CUM TOTAL	CUM AVG	UNIT
1	1.00000000	1.00000000	1.00000000	1.00000000	1.00000000	1.00000000
2	1.89000000	.94500000	.89000000	1.90000000	.95000000	.90000000
3	2.72135222	.90711741	.83135222	2.74620599	.91540200	.84620599
4	3.51345222	.87836305	.79210000	3.55620599	.88905150	.81000000
5	4.27638668	.85527734	.76293447	4.33919271	.86783854	.78298672
6	5.01629015	.83604836	.73990347	5.10077810	.85012968	.76158539
7	5.73726446	.81960921	.72097430	5.84472593	.83496085	.74394783
8	6.44223346	.80527918	.70496900	6.57372593	.82171574	.72900000
9	7.13337996	.79259777	.69114651	7.28979050	.80997672	.71606457
10	7.81239164	.78123916	.67901167	7.99447855	.79944786	.70468805
11	8.48060966	.77096451	.66821803	8.68903107	.78991192	.69455252
12	9.13912375	.76159365	.65851409	9.37445792	.78120483	.68542685
13	9.78883558	.75298735	.64971182	10.05159588	.77319968	.67713797
14	10.43050271	.74503591	.64166713	10.72114893	.76579635	.66955305
15	11.06476997	.73765133	.63426726	11.38371699	.75891447	.66256805
16	11.69219238	.73076202	.62742241	12.03981699	.75248856	.65610000
17	12.31325234	.72430896	.62105996	12.68989871	.74646463	.65008173
18	12.92837273	.71824293	.61512039	13.33435682	.74079760	.64445811
19	13.53792705	.71252248	.60955432	13.97354024	.73544949	.63918341
20	14.14224744	.70711237	.60432039	14.60775948	.73038797	.63421924
21	14.74163103	.70198243	.59938358	15.23729259	.72558536	.62953311
22	15.33634507	.69710659	.59471405	15.86238986	.72101772	.62509727
23	15.92663118	.69246223	.59028610	16.48327770	.71666425	.62088784
24	16.51270872	.68802953	.58607754	17.10016187	.71250674	.61688416
25	17.09477772	.68379111	.58206900	17.71323007	.70852920	.61306821
26	17.67302124	.67973159	.57824352	18.32265424	.70471747	.60942417
27	18.24760742	.67583731	.57458618	18.92859237	.70105898	.60593813
28	18.81869116	.67209611	.57108375	19.53119012	.69754250	.60259775
29	19.38641563	.66849709	.56772447	20.13058217	.69415801	.59939205
30	19.95091349	.66503045	.56449786	20.72689341	.69089645	.59631125
31	20.51230800	.66168735	.56139451	21.32023995	.68774968	.59334653
32	21.07071394	.65845981	.55840594	21.91072995	.68471031	.59049000
33	21.62623848	.65534056	.55552454	22.49864645	.68177165	.58773450
34	22.17898185	.65232300	.55274337	23.08353800	.67892759	.58507355
35	22.72903799	.64940109	.55005614	23.66603927	.67617255	.58250128
36	23.27649514	.64656931	.54745715	24.24605158	.67350143	.58001230
37	23.82143629	.64382260	.54494114	24.82365331	.67090955	.57760173
38	24.36393963	.64115631	.54250335	25.39891838	.66839259	.57526507
39	24.90407900	.63856613	.54013936	25.97191658	.66594658	.57299820
40	25.44192415	.63604810	.53784515	26.54271390	.66356785	.57079732
41	25.97754112	.63359856	.53561697	27.11137283	.66125300	.56865893
42	26.51099251	.63121411	.53345139	27.67795263	.65899887	.56657980
43	27.04233772	.62889157	.53134522	28.24250956	.65680255	.56455693
44	27.57163323	.62662803	.52929550	28.80509710	.65466130	.56258754
45	28.09893272	.62442073	.52729949	29.36576615	.65257258	.56066905
46	28.62428735	.62226712	.52535463	29.92456521	.65053403	.55879906
47	29.14774589	.62016481	.52345855	30.48154053	.64854342	.55697532
48	29.66935490	.61811156	.52160901	31.03673628	.64659867	.55519575
49	30.18915885	.61610528	.51980395	31.59019466	.64469785	.55345838
50	30.70720026	.61414401	.51804141	32.14195605	.64283912	.55176139

LEARNING CURVE

Percent Learning:		89.00%			90.00%	
N	CUM	CUM AVG	UNIT	CUM TOTAL	CUM AVG	UNIT
51	31.22351984	.61222588	.51631958	32.69205909	.64102077	.55010305
52	31.73815657	.61034916	.51463674	33.24054085	.63924117	.54848175
53	32.25114785	.60851222	.51299128	33.78743683	.63749881	.54689598
54	32.76252955	.60671351	.51138170	34.33278114	.63579224	.54534431
55	33.27233612	.60495157	.50980656	34.87660654	.63412012	.54382540
56	33.78060065	.60322501	.50826453	35.41894451	.63248115	.54233797
57	34.28735499	.60153254	.50675434	35.95982535	.63087413	.54088083
58	34.79262977	.59987293	.50527478	36.49927819	.62929790	.53945285
59	35.29645449	.59824499	.50382472	37.03733114	.62775138	.53805295
60	35.79885758	.59664763	.50240310	37.57401126	.62623352	.53668012
61	36.29986646	.59507978	.50100888	38.10934467	.62474336	.53533340
62	36.79950757	.59354044	.49964111	38.64335655	.62327994	.53401188
63	37.29780644	.59202867	.49829887	39.17607123	.62184240	.53271469
64	37.79478773	.59054356	.49698129	39.70751223	.62042988	.53144100
65	38.29047528	.58908424	.49568754	40.23770227	.61904157	.53019004
66	38.78489212	.58764988	.49441684	40.76666332	.61767672	.52896105
67	39.27806055	.58623971	.49316843	41.29441665	.61633458	.52775333
68	39.77000214	.58485297	.49194160	41.82098284	.61501445	.52656620
69	40.26073780	.58348895	.49073566	42.34638185	.61371568	.52539901
70	40.75028777	.58214697	.48954997	42.87063300	.61243761	.52425115
71	41.23867167	.58082636	.48838390	43.39375502	.61117965	.52312202
72	41.72590853	.57952651	.48723686	43.91576610	.60994120	.52201107
73	42.21201681	.57824681	.48610828	44.43668385	.60872170	.52091776
74	42.69701443	.57698668	.48499762	44.95652541	.60752061	.51984156
75	43.18091878	.57574558	.48390435	45.47530740	.60633743	.51878199
76	43.66374676	.57452298	.48282798	45.99304596	.60517166	.51773857
77	44.14551479	.57331837	.48176803	46.50975680	.60402282	.51671084
78	44.62623883	.57213127	.48072403	47.02545518	.60289045	.51569838
79	45.10593439	.57096119	.47969556	47.54015595	.60177413	.51470076
80	45.58461657	.56980771	.47868218	48.05387354	.60067342	.51371759
81	46.06230006	.56867037	.47768349	48.56662201	.59958793	.51274847
82	46.53899916	.56754877	.47669910	49.07841504	.59851726	.51179304
83	47.01472780	.56644250	.47572864	49.58926598	.59746104	.51085094
84	47.48949954	.56535118	.47477174	50.09918780	.59641890	.50992182
85	47.96332759	.56427444	.47382805	50.60819316	.59539051	.50900536
86	48.43622483	.56321192	.47289724	51.11629439	.59437552	.50810124
87	48.90820383	.56216326	.47197900	51.62350354	.59337360	.50720914
88	49.37927683	.56112815	.47107300	52.12983232	.59238446	.50632879
89	49.84945577	.56010624	.47017894	52.63529220	.59140778	.50545988
90	50.31875231	.55909725	.46929655	53.13989435	.59044327	.50460215
91	50.78717784	.55810086	.46842553	53.64364967	.58949066	.50375532
92	51.25474347	.55711678	.46756562	54.14656882	.58854966	.50291915
93	51.72146003	.55614473	.46671657	54.64866221	.58762002	.50209339
94	52.18733814	.55518445	.46587811	55.14994000	.58670149	.50127779
95	52.65238814	.55423566	.46505000	55.65041213	.58579381	.50047213
96	53.11662016	.55329813	.46423202	56.15008830	.58489675	.49967617
97	53.58004409	.55237159	.46342393	56.64897802	.58401008	.49888971
98	54.04266960	.55145581	.46262551	57.14709056	.58313358	.49811254
99	54.50450616	.55055057	.46183656	57.64443501	.58226702	.49734445
100	54.96556301	.54965563	.46105685	58.14102026	.58141020	.49658525

LEARNING CURVE

Percent Learning:	91.00%			92.00%		
N	CUM TOTAL	CUM AVG	UNIT	CUM TOTAL	CUM AVG	UNIT
1	1.00000000	1.00000000	1.00000000	1.00000000	1.00000000	1.00000000
2	1.91000000	.95500000	.91000000	1.92000000	.96000000	.92000000
3	2.77115662	.92371887	.86115662	2.79620366	.93206789	.87620366
4	3.59925662	.89981415	.82810000	3.64260366	.91065092	.84640000
5	4.40259230	.88051846	.80333568	4.46658607	.89331721	.82398241
6	5.18624482	.86437414	.78365252	5.27269344	.87878224	.80610737
7	5.95363220	.85051889	.76738738	6.06399055	.86628436	.79129711
8	6.70720320	.83840040	.75357100	6.84267855	.85533482	.77868800
9	7.44879392	.82764377	.74159072	7.61041141	.84560127	.76773286
10	8.17982939	.81798294	.73103547	8.36847522	.83684752	.75806381
11	8.90144596	.80922236	.72161657	9.11789727	.82889975	.74942205
12	9.61456975	.80121415	.71312379	9.85951605	.82162634	.74161878
13	10.31996924	.79384379	.70539949	10.59402829	.81492525	.73451224
14	11.01829176	.78702084	.69832251	11.32202163	.80871583	.72799334
15	11.71008959	.78067264	.69179784	12.04399803	.80293320	.72197640
16	12.39583920	.77473995	.68574961	12.76039099	.79752444	.71639296
17	13.07595555	.76917386	.68011635	13.47157845	.79244579	.71118746
18	13.75080310	.76393351	.67484755	14.17789268	.78766070	.70631423
19	14.42070438	.75898444	.66990127	14.87962797	.78313831	.70173529
20	15.08594666	.75429733	.66524228	15.57704668	.77885233	.69741871
21	15.74678737	.74984702	.66084072	16.27038410	.77478020	.69333742
22	16.40345845	.74561175	.65667108	16.95985238	.77090238	.68946828
23	17.05616986	.74157260	.65271140	17.64564372	.76720190	.68579133
24	17.70511251	.73771302	.64894265	18.32793300	.76366387	.68228928
25	18.35046073	.73401843	.64534822	19.00688000	.76027520	.67894701
26	18.99237427	.73047593	.64191354	19.68263126	.75702428	.67575126
27	19.63100002	.72707407	.63862575	20.35532161	.75390080	.67269034
28	20.26647351	.72380263	.63547349	21.02507548	.75089555	.66975387
29	20.89892011	.72065242	.63244660	21.69200809	.74800028	.66693261
30	21.52845614	.71761520	.62953603	22.35622638	.74520755	.66421829
31	22.15518980	.71468354	.62673366	23.01782987	.74251064	.66160349
32	22.77922194	.71185069	.62403215	23.67691139	.73990348	.65908152
33	23.40064682	.70911051	.62142488	24.33355774	.73738054	.65664634
34	24.01955270	.70645743	.61890588	24.98785020	.73493677	.65429246
35	24.63602236	.70388635	.61646966	25.63986509	.73256757	.65201490
36	25.25013364	.70139260	.61411127	26.28967419	.73026873	.64980909
37	25.86195980	.69897189	.61182616	26.93734507	.72803635	.64767089
38	26.47156996	.69662026	.60961016	27.58294154	.72586688	.64559646
39	27.07902939	.69433409	.60745944	28.22652385	.72375702	.64358232
40	27.68439987	.69211000	.60537007	28.86814906	.72170373	.64162521
41	28.28773988	.68994488	.60334001	29.50787123	.71970418	.63972217
42	28.88910494	.68783583	.60136505	30.14574167	.71775575	.63787043
43	29.48854774	.68578018	.59944280	30.78180910	.71585603	.63606744
44	30.08611842	.68377542	.59757068	31.41611992	.71400273	.63431082
45	30.68186471	.68181922	.59574629	32.04871829	.71219374	.63259837
46	31.27583209	.67990939	.59396738	32.67964632	.71042709	.63092803
47	31.86806396	.67804391	.59223187	33.30894420	.70870094	.62929788
48	32.45860177	.67622087	.59053781	33.93665034	.70701355	.62770614
49	33.04748516	.67443847	.58888339	34.56280145	.70536329	.62615111
50	33.63475204	.67269504	.58726688	35.18743270	.70374865	.62463125

LEARNING CURVE

Percent Learning:		91.00%				92.00%		
N	CUM TOTAL	CUM AVG	UNIT		CUM TOTAL	CUM AVG	UNIT	
51	34.22043873	.67098899	.58568669		35.81057775	.70216819	.62314506	
52	34.80458004	.66931885	.58414132		36.43226892	.70062056	.62169116	
53	35.38720939	.66768320	.58262934		37.05253717	.69910447	.62026826	
54	35.96835882	.66608072	.58114944		37.67141229	.69761875	.61887512	
55	36.54805916	.66451017	.57970034		38.28892287	.69616223	.61751058	
56	37.12634004	.66297036	.57828087		38.90509643	.69473386	.61617356	
57	37.70322995	.66146017	.57688991		39.51995946	.69333262	.61486303	
58	38.27875636	.65997856	.57552641		40.13353746	.69195754	.61357800	
59	38.85294571	.65852450	.57418935		40.74585502	.69060771	.61231756	
60	39.42582350	.65709706	.57287779		41.35693585	.68928226	.61108083	
61	39.99741433	.65569532	.57159083		41.96680282	.68798037	.60986697	
62	40.56774196	.65431842	.57032763		42.57547803	.68670126	.60867521	
63	41.13682931	.65296554	.56908736		43.18298282	.68544417	.60750479	
64	41.70469857	.65163592	.56786925		43.78933783	.68420840	.60635500	
65	42.27137115	.65032879	.56667258		44.39456299	.68299328	.60522516	
66	42.83686779	.64904345	.56549664		44.99867762	.68179815	.60411463	
67	43.40120857	.64777923	.56434078		45.60170042	.68062239	.60302280	
68	43.96441292	.64653548	.56320435		46.20364949	.67946543	.60194907	
69	44.52649966	.64531159	.56208674		46.80454237	.67832670	.60089288	
70	45.08748705	.64410696	.56098739		47.40439607	.67720566	.59985370	
71	45.64739279	.64292103	.55990574		48.00322710	.67610179	.59883103	
72	46.20623405	.64175325	.55884126		48.60105146	.67501460	.59782436	
73	46.76402750	.64060312	.55779344		49.19788471	.67394363	.59683324	
74	47.32078930	.63947013	.55676181		49.79374192	.67288840	.59585722	
75	47.87653519	.63835380	.55574589		50.38863778	.67184850	.59489585	
76	48.43128044	.63725369	.55474524		50.98258652	.67082351	.59394875	
77	48.98503988	.63616935	.55375945		51.57560202	.66981301	.59301550	
78	49.53782797	.63510036	.55278809		52.16769775	.66881664	.59209573	
79	50.08965875	.63404631	.55183078		52.75888683	.66783401	.59118908	
80	50.64054588	.63300682	.55088713		53.34918203	.66686478	.59029520	
81	51.19050267	.63198151	.54995679		53.93859577	.66590859	.58941374	
82	51.73954209	.63097003	.54903941		54.52714016	.66496512	.58854440	
83	52.28767674	.62997201	.54813466		55.11482701	.66403406	.58768685	
84	52.83491894	.62898713	.54724220		55.70166781	.66311509	.58684080	
85	53.38128067	.62801507	.54636173		56.28767376	.66220793	.58600595	
86	53.92677362	.62705551	.54549295		56.87285581	.66131228	.58518204	
87	54.47140919	.62610815	.54463557		57.45722460	.66042787	.58436880	
88	55.01519852	.62517271	.54378932		58.04079056	.65955444	.58356595	
89	55.55815244	.62424890	.54295392		58.62356383	.65869173	.58277327	
90	56.10028156	.62333646	.54212912		59.20555433	.65783949	.58199050	
91	56.64159622	.62243512	.54131466		59.78677174	.65699749	.58121741	
92	57.18210654	.62154464	.54051031		60.36722552	.65616549	.58045379	
93	57.72182237	.62066476	.53971583		60.94692493	.65534328	.57969940	
94	58.26075337	.61979525	.53893100		61.52587898	.65453063	.57895405	
95	58.79890897	.61893588	.53815560		62.10409651	.65372733	.57821753	
96	59.33629838	.61808644	.53738941		62.68158615	.65293319	.57748964	
97	59.87293062	.61724671	.53663224		63.25835636	.65214800	.57677020	
98	60.40881450	.61641647	.53588388		63.83441538	.65137159	.57605903	
99	60.94395865	.61559554	.53514415		64.40977131	.65060375	.57535593	
100	61.47837151	.61478372	.53441286		64.98443206	.64984432	.57466075	

LEARNING CURVE

Percent Learning:		93.00%				94.00%	
N	CUM TOTAL	CUM AVG	UNIT	CUM TOTAL	CUM AVG	UNIT
1	1.00000000	1.00000000	1.00000000	1.00000000	1.00000000	1.00000000
2	1.93000000	.96500000	.93000000	1.94000000	.97000000	.94000000
3	2.82134669	.94044890	.89134669	2.84658526	.94886175	.90658526
4	3.68624669	.92156167	.86490000	3.73018526	.93254632	.88360000
5	4.53117463	.90623493	.84492794	4.59635860	.91927172	.86617333
6	5.36012705	.89335451	.82895242	5.44854874	.90809146	.85219015
7	6.17580825	.88225832	.81568120	6.28909254	.89844179	.84054380
8	6.98016525	.87252066	.80435700	7.11967654	.88995957	.83058400
9	7.77466417	.86385157	.79449892	7.94157338	.88239704	.82189684
10	8.56044715	.85604472	.78578299	8.75577631	.87557763	.81420293
11	9.33842802	.84894800	.77798087	9.56308132	.86937103	.80730501
12	10.10935377	.84244615	.77092575	10.36414006	.86367834	.80105874
13	10.87384596	.83644969	.76449219	11.15949548	.85842273	.79535541
14	11.63242947	.83088782	.75858351	11.94960665	.85354333	.79011117
15	12.38555319	.82570355	.75312372	12.73486663	.84899111	.78525998
16	13.13360520	.82085033	.74805201	13.51561559	.84472597	.78074896
17	13.87692418	.81628966	.74331898	14.29215070	.84071475	.77653512
18	14.61580818	.81198934	.73888399	15.06473373	.83692965	.77258303
19	15.35052139	.80792218	.73471321	15.83359692	.83334721	.76886319
20	16.08129957	.80406498	.73077818	16.59894768	.82994738	.76535076
21	16.80835430	.80039782	.72705473	17.36097230	.82671297	.76202462
22	17.53187651	.79690348	.72352221	18.11983901	.82362905	.75886671
23	18.25203927	.79356692	.72016277	18.87570043	.82068263	.75586142
24	18.96900022	.79037501	.71696095	19.62869565	.81786232	.75299522
25	19.68290345	.78731614	.71390323	20.37895189	.81515808	.75025624
26	20.39388118	.78438005	.71097773	21.12658598	.81256100	.74763409
27	21.10205516	.78155760	.70817398	21.87170554	.81006317	.74511956
28	21.80753783	.77884064	.70548267	22.61441004	.80765750	.74270450
29	22.51043333	.77622184	.70289550	23.35479165	.80533764	.74038161
30	23.21083839	.77369461	.70040506	24.09293603	.80309787	.73814438
31	23.90884308	.77125300	.69800469	24.82892298	.80093300	.73598695
32	24.60453145	.76889161	.69568837	25.56282700	.79883834	.73390402
33	25.29798212	.76660552	.69345067	26.29471783	.79680963	.73189082
34	25.98926877	.76439026	.69128665	27.02466083	.79484297	.72994301
35	26.67846061	.76224173	.68919183	27.75271746	.79293478	.72805662
36	27.36562272	.76015619	.68716212	28.47894550	.79108182	.72622805
37	28.05081647	.75813017	.68519375	29.20339949	.78928107	.72445399
38	28.73409976	.75616052	.68328329	29.92613089	.78752976	.72273140
39	29.41552733	.75424429	.68142754	30.64718839	.78582534	.72105750
40	30.09515104	.75237878	.67962370	31.36661810	.78416545	.71942971
41	30.77302001	.75056146	.67786898	32.08446376	.78254790	.71784566
42	31.44918092	.74879002	.67616090	32.80076064	.78097064	.71630314
43	32.12367809	.74706228	.67449717	33.51556703	.77943179	.71480012
44	32.79655374	.74537622	.67287565	34.22890173	.77792958	.71333471
45	33.46784808	.74372996	.67129434	34.94080686	.77646237	.71190513
46	34.13759945	.74212173	.66975137	35.65131659	.77502862	.71050974
47	34.80584448	.74054988	.66824503	36.36046360	.77362689	.70914701
48	35.47261816	.73901288	.66677368	37.06827911	.77225581	.70781550
49	36.13795398	.73750926	.66533581	37.77479298	.77091414	.70651387
50	36.80188398	.73603768	.66393000	38.48003385	.76960068	.70524087

LEARNING CURVE

Percent Learning:	93.00%			94.00%		
N	CUM TOTAL	CUM AVG	UNIT	CUM TOTAL	CUM AVG	UNIT
51	37.46443889	.73459684	.66255491	39.18402914	.76831430	.70399529
52	38.12564818	.73318554	.66120929	39.88680518	.76705395	.70277604
53	38.78554014	.73180264	.65989196	40.58838726	.76581863	.70158207
54	39.44414195	.73044707	.65860180	41.28879965	.76460740	.70041239
55	40.10147972	.72911781	.65733777	41.98806572	.76341938	.69926607
56	40.75757860	.72781390	.65609888	42.68620795	.76225371	.69814223
57	41.41246279	.72653443	.65488419	43.38324799	.76110961	.69704004
58	42.06615560	.72527854	.65369282	44.07920670	.75998632	.69595871
59	42.71867952	.72404542	.65252392	44.77410421	.75888312	.69489751
60	43.37005623	.72283427	.65137671	45.46795993	.75779933	.69385572
61	44.02030666	.72164437	.65025043	46.16079260	.75673430	.69283267
62	44.66945102	.72047502	.64914436	46.85262032	.75568742	.69182773
63	45.31750885	.71932554	.64805783	47.54346062	.75465811	.69084029
64	45.96449903	.71819530	.64699018	48.23333040	.75364579	.68986978
65	46.61043984	.71708369	.64594081	48.92224605	.75264994	.68891565
66	47.25534896	.71599014	.64490912	49.61022342	.75167005	.68797738
67	47.89924352	.71491408	.64389456	50.29727788	.75070564	.68705446
68	48.54214011	.71385500	.64289659	50.98342431	.74975624	.68614643
69	49.18405481	.71281239	.64191470	51.66867714	.74882141	.68525283
70	49.82500321	.71178576	.64094841	52.35305036	.74790072	.68437322
71	50.46500046	.71077465	.63999724	53.03655757	.74699377	.68350720
72	51.10406122	.70977863	.63906077	53.71921193	.74610017	.68265437
73	51.74219977	.70879726	.63813855	54.40102627	.74521954	.68181434
74	52.37942996	.70783013	.63723018	55.08201301	.74435153	.68098675
75	53.01576523	.70687687	.63633528	55.76218427	.74349579	.68017125
76	53.65121869	.70593709	.63545346	56.44155178	.74265200	.67936752
77	54.28580305	.70501043	.63458436	57.12012700	.74181983	.67857522
78	54.91953070	.70409655	.63372765	57.79792105	.74099899	.67779405
79	55.55241368	.70319511	.63288298	58.47494476	.74018917	.67702371
80	56.18446373	.70230580	.63205005	59.15120869	.73939011	.67626393
81	56.81569226	.70142830	.63122853	59.82672311	.73860152	.67551442
82	57.44611041	.70056232	.63041815	60.50149803	.73782315	.67477492
83	58.07572902	.69970758	.62961861	61.17554321	.73705474	.67404518
84	58.70455866	.69886379	.62882964	61.84886816	.73629605	.67332495
85	59.33260963	.69803070	.62805098	62.52148217	.73554685	.67261401
86	59.95989201	.69720805	.62728237	63.19339429	.73480691	.67191212
87	60.58641558	.69639558	.62652358	63.86461334	.73407602	.67121906
88	61.21218994	.69559307	.62577436	64.53514797	.73335395	.67053462
89	61.83722442	.69480027	.62503448	65.20500658	.73264052	.66985861
90	62.46152815	.69401698	.62430373	65.87419739	.73193553	.66919082
91	63.08511005	.69324297	.62358190	66.54272845	.73123877	.66853106
92	63.70797883	.69247803	.62286878	67.21060761	.73055008	.66787915
93	64.33014300	.69172197	.62216417	67.87784253	.72986927	.66723492
94	64.95161088	.69097458	.62146788	68.54444072	.72919618	.66659819
95	65.57239060	.69023569	.62077972	69.21040951	.72853063	.66596879
96	66.19249012	.68950511	.62009952	69.87575608	.72787246	.66534657
97	66.81191723	.68878265	.61942711	70.54048745	.72722152	.66473137
98	67.43067954	.68806816	.61876231	71.20461049	.72657766	.66412304
99	68.04878449	.68736146	.61810496	71.86813193	.72594073	.66352144
100	68.66623940	.68666239	.61745490	72.53105835	.72531058	.66292641

Appendix B
Tables of Construction Job Factors

TABLE B-1 Size of construction job factor.

From	To	Factor
$ 0	$ 99,999	0.120
100,000	199,999	0.119
200,000	299,999	0.117
300,000	399,999	0.116
400,000	499,999	0.114
500,000	599,999	0.113
600,000	699,999	0.111
700,000	799,999	0.110
800,000	899,999	0.109
900,000	999,999	0.107
1,000,000	1,099,999	0.106
1,100,000	1,199,999	0.104
1,200,000	1,299,999	0.103
1,300,000	1,399,999	0.101
1,400,000	1,499,999	0.100
1,500,000	1,599,999	0.099
1,600,000	1,699,999	0.097
1,700,000	1,799,999	0.096
1,800,000	1,899,999	0.094
1,900,000	1,999,999	0.093
2,000,000	2,099,999	0.091
2,100,000	2,199,999	0.090
2,200,000	2,299,999	0.089
2,300,000	2,399,999	0.087
2,400,000	2,499,999	0.086

TABLE B-1 Size of construction job factor. (*continued*)

From	To	Factor
2,500,000	2,599,999	0.084
2,600,000	2,699,999	0.083
2,700,000	2,799,999	0.081
2,800,000	2,899,999	0.080
2,900,000	2,999,999	0.079
3,000,000	3,099,999	0.077
3,100,000	3,199,999	0.076
3,200,000	3,299,999	0.074
3,300,000	3,399,999	0.073
3,400,000	3,499,999	0.071
3,500,000	3,599,999	0.070
3,600,000	3,699,999	0.069
3,700,000	3,799,999	0.067
3,800,000	3,899,999	0.066
3,900,000	3,999,999	0.064
4,000,000	4,099,999	0.063
4,100,000	4,199,999	0.061
4,200,000	4,299,999	0.060
4,300,000	4,399,999	0.059
4,400,000	4,499,999	0.057
4,500,000	4,599,999	0.056
4,600,000	4,699,999	0.054
4,700,000	4,799,999	0.053
4,800,000	4,899,999	0.051
4,900,000	4,999,999	0.050
5,000,000	9,999,999	0.040
10,000,000	more than 10,000,000	0.030

Tables of Construction Job Factors 201

TABLE B-2 Period of construction job performance factor.

Period of Performance	Factor
Over 24 months	0.120
23 to 24 months	0.116
22 to 23 months	0.112
21 to 22 months	0.109
20 to 21 months	0.105
19 to 20 months	0.101
18 to 19 months	0.098
17 to 18 months	0.094
16 to 17 months	0.090
15 to 16 months	0.086
14 to 15 months	0.082
13 to 14 months	0.079
12 to 13 months	0.075
11 to 12 months	0.071
10 to 11 months	0.068
9 to 10 months	0.064
8 to 9 months	0.060
7 to 8 months	0.056
6 to 7 months	0.052
5 to 6 months	0.049
4 to 5 months	0.045
3 to 4 months	0.041
2 to 3 months	0.038
1 to 2 months	0.034
UNDER 30 days	0.030

TABLE B-3 Construction subcontracting factor.

Subcontracting	Factor
80% or more	0.030
70% to 79.9%	0.042
60% to 69.9%	0.055
50% to 59.9%	0.068
40% to 49.9%	0.080
30% to 39.9%	0.092
20% to 29.9%	0.105
10% to 19.9%	0.118
0% to 9.9%	0.120

Appendix C
Sample Format for Postnegotiation Summary

NEGOTIATED COSTS/PRICE

Cost Element	Estimate	Supplier's Proposal(a)	Cost/Price Analyst's Position(b)	Negotiator's Objectives(c)	Negotiated (d)	Difference c and d	Remarks
Material							
Direct Labor							
Labor Overhead ($ / %)							
Other Direct Costs							
TOTAL DIRECT COST AND OVERHEAD							
General Admin. Expense ($ / %) .							
TOTAL EST. COST							
Fee or Profit ($ / %)							
TOTAL COST & FEE OR PROFIT							

Figure C-1. Sample format for postnegotiation summary.

Index

ABC Analysis, 167
Accounting analysis, 95, 111–127. *See also* Rate analysis
Additional inventory carrying cost, 139–140
Adequate price competition, 169. *See also* Effective competition; Price competition
Administrative costs (of purchasing), 47
Aesthetic qualities, 46. *See also* Utility qualities; Value Analysis
Agreements, 32
All or none, 47–49. *See also* Any or all
Allocable cost, 169
Allowable cost, 169
  incurred, 26
Alternate Dispute Resolution (ADR), 142. *See also* Mediation and arbitration
American Management Association, 111
Any or all, 48–49. *See also* All or none
Arbitration, 142. *See also* Alternate dispute resolution (ADR)
"Area Wage Surveys," 38
Assist, 138
Association of General Contractors, 118
Audit, 2, 119
Auditors, 2

Authorization to proceed, 156–158
Auxiliary price analysis technique, 46. *See also* Tertiary price analysis technique
Award fee, 28. *See also* Base fee

Base Amount, 17–20. *See also* Guaranteed minimum quantity
Base fee, 28. *See also* Award fee
Bottom-up estimate, 5. *See also* Estimated costs; In-house estimate
Break-even point, 73. *See also* Breakeven
Breakeven, 72–73. *See also* Break-even point
  volume, 72–73
  sales price, 72–73
Brick and mortar cost, 63. *See also* Direct cost
Burden, 2, 63. *See also* Costs of doing business; Indirect cost; Overhead

Catalog or market prices of commercial items sold to the general public in substantial quantities, 6
Ceiling or cap 17. *See also* Ceiling price

205

Index

Ceiling price, 14–17. *See also* Ceiling or cap
Certificate of inspection, 139
Change order, 156–158. *See also* Contract modification
   unilateral, 156
Changes clause, 156–157
Changes:
   in delivery and order quantity requirements, 166
   in inventory and materials management, 167
   in procurement and buying methods, 167
   in quality or design of the products and services purchased, 165–166
   in supplier base, 166
Combination contracts, 30–31. *See also* Composite contracts
Comparison, *see* Price analysis
Competitive bidding, 11
Composite contracts, 30–31. *See also* Combination contracts
Consolidated bill of materials, 110
Consumer price index, 38. *See also* Producer price index
Contract modification, 156–158. *See also* Change order
Convention on Contracts for the International Sale of Goods (CISG), 142
Cost analysis, 1–3, 95–127, 169
Cost analytical methods, 1
Cost behavior, 64
Cost estimating, 79
   round-table, 79–80
   comparison, 80
   detailed, 81–86
   combination of methods, 82
Cost estimators, 2
Cost incentive, 27. *See also* Performance incentive
Cost no-fee/cost sharing, 26
Cost or pricing data, 86. *See also* Current, accurate, and complete; Federal norm; Truth-in-Negotiations Act
Cost plus Insurance and Freight (CIF), 138–139
Cost reimbursement contract, 9–12, 36–30, 170. *See also* Cost-plus
Cost-plus:
   a-percentage-of-cost, 30
   award-fee, 28–30
   fixed-fee, 27
   incentive-fee, 27–29
Cost-volume-profit relationship, 64
Cost/price analysts, 2
Costs of doing business, 63. *See also* Burden; Indirect cost; Overhead
Cumulative average (learning) curve, 97. *See also* Unit learning curve
"Current Wage Developments," 38
Current, accurate, and complete, 86. *See also* Cost or pricing data; Federal norm, Truth-in-Negotiations Act
Customs duties, 139
Customshouse broker fee, 139

Dataquest, 118
Delivery orders, 23
   order for supplies or services, 23
   letter format, 24
Demurrage, 139
Dependent variable, 41–43. *See also* Independent variable
Depreciation, method (s) of:
   sum-of-the-years digits, 123–127
   declining-balance, 123–127
   straight line, 124–127
   service hour, 124–127
Direct cost, 63–64, 170. *See also* Brick and mortar cost
   direct labor, 63–64, 170
   direct material, 63
   other direct, 63
Discount:
   trade (favored customer), 49–51
   term (prompt payment), 49–51

Index   207

Documentation, 145–150

Effective competition, 6–7. *See also* Adequate price competition; Price competition
Engineering analysis, 95–111. *See also* Quantitative analysis
EOQ, 167
Escalation cost, 138
Established catalog price, 170
Established market price, 170
Estimated costs, 3, 79–86. *See also* Bottom-up estimate; In-house estimate
Experience curve, 97. *See also* Improvement curve; Learning curve
Export processing fee, 139
Extra communication and documentation cost, 139–140

F.O.B. (Free On Board)
    origin, 52, 143
    origin, freight allowed, 52–53
    destination. 53
Factor weight, 130–133
Fair and reasonable profit or fee, 129
Federal Norm, 86. *See also* Cost or pricing data; Current, accurate, and complete; Truth-in-Negotiations
Fee adjustment ratio, 27–28. *See also* Share ratio
Fee, 27–29, 171
    maximum, 27–29
    minimum, 27–29
Financing charge, 139
Firm-fixed-price contract, 9–12
Fixed cost per unit, 66–72
Fixed cost, 64–66, 171
Fixed price contract, 9–17, 171
Fixed-price incentive, 13–17. *See also* Performance Incentive

Fixed-price with economic price adjustment, 11–13
    adjustments based on actual costs of labor or material, 13
    adjustments based on cost indexes of labor or material, 13
    adjustments based on established prices, 13
Flexibly-priced contract, 9, 11
Foreign exchange conversion, 139

General and Administrative (G&A) rate, 119. *See also* General and Administrative (G&A); Overhead
Guaranteed minimum quantity, 17–20, 23. *See also* Base amount

Hedging, 139
    forward contract, 139
    futures contract, 139
    currency or commodity future, 139
Historical cost data, 3
Hybrid contracts, 30
    time-and materials, 30
    labor-hour, 30

Improvement curve, 97. *See also* Experience curve; Learning curve
In-house estimate, 5–6, 79–86. *See also* Bottom-up estimate; Estimated cost
Incentive contract, 9
Incoterms, 53–54
Indefinite-delivery type, 17–26. *See also* Open end; Term
    requirements, 17, 20–26
    indefinite-quantity indefinite delivery, 17–26
    definite-quantity indefinite delivery, 17
Independent variable, 41–43. *See also* Dependent variable

Indirect cost, 63–64, 171. *See also* Burden; Costs of doing business; Overhead
Inland transportation cost, 139
International Chamber of Commerce (ICC), 53

Judgmental factors, 3
Just-In-Time, 166

Known-known, 142. *See also* Known-unknown and Unknown-unknown
Known-unknown, 142. *See also* Known-known and Unknown-unknown

Labor rate, 2, 111–116
Learning curve, 97–101. *See also* Experience curve; Improvement curve
Learning phenomenon, 97
Letter contract, 32
Letter of credit, 139
Leverage, 70
Line of best fit, 43
Litigation, 142, 145
Log-log graph paper, 98. *See also* Logarithmic scale
Logarithmic scale, 98. *See also* Log-log graph paper

Make-or-buy analysis, 167
Make-or-buy, 5, 167, 172
Marine insurance premium, 139
Material handling costs, 30
Maximum contract value, 23
Maximum order, 23
Maximum quantity, 17
Mediation, 142. *See also* Alternate Dispute Resolution (ADR)
Memorandum of price negotiation, 145. *See also* Negotiation memorandum; Price/negotiation memorandum (a)
Minimum order, 23
Modified commercial item, 137–138
"Monthly Labor Review," 38
Most favored customers, 34
MRO material, 166

Negotiation memorandum. 145, 161–163. *See also* Memorandum of price negotiation; Price/Negotiation memorandum (a)
Negotiation objective, 145–150
Negotiation, 145–158
Non-Federal Norm, 86

Ocean shipping cost, 139
Off-shore source, 138
Oligopolistic industry, 6
Open end, 17. *See also* Indefinite-delivery type and Term
Other direct cost, 110, 118–119
Other than price-related factors, 11. *See also* Price-related factors
Out-of-scope change, 46
Overhead 2, 63. *See also* Burden; Costs of doing business; Indirect cost; Overhead rate
  material, 119–120
  manufacturing, 119–123
  General and Administrative (G&A), 119
Overhead and G&A pool, 119
Overhead rate, 119

Performance incentive, 27. *See also* Cost incentive
Piggyback, 53
Point of total assumption, 15–17
Port handling cost, 139
Port of debarkation, 139

Index   209

Post-negotiation summary, 162
Pre-negotiation documentation, 150
Prepay and add, 53
Price analysis, 1–8, 33–60, 171. *See also* Price analytical/comparison techniques
   with incremental cost analysis, 4–5, 137–143. *See also* Primary reliance on price analysis; Secondary reliance on cost analysis
   independent estimates of cost, 3
   market data (indexes), 3
   other prices and quotations submitted, 2
   prices for the same or similar items, 3
   prices set by law or regulation, 3
   prior quotations for the same or similar items, 3
   published catalog or market prices, 2
   rough yardsticks, 3
   value analysis, 3
   visual analysis, 3
Price analytical/comparison techniques, 1. *See also* Price analysis
Price and cost analysis, 4–5
Price competition, 6. *See also* Adequate price competition; Effective competition
Price-related factors, 11. *See also* Other than price-related factors
Price/negotiation memorandum (a), 2. *See also* Memorandum of price negotiation; Negotiation Memorandum
Pricing decision, 5
Pricing Team, 1–2, 5, 95, 147, 149–151
Primary method of price analysis, 34
Primary reliance on price analysis, 137. See also Price analysis with incremental cost analysis; Secondary reliance on cost analysis
Procurement method, 171

Producer price index, 38–39. *See also* Consumer price index
Profit: 172
   analysis, 1, 129–133
   or fee determinant, 129
Profit position, 1
Prospective pricing, 9, 119, 172

Quantitative analysis, 96. *See also* Engineering analysis

Range of Incentive Effectiveness (RIE), 27–29
Rate analysis, 111. *See also* Accounting analysis
Rate of learning, 98. *See also* Slope of the learning curve
Reasonable economy and efficiency, 3
Reasonableness, 172
Regulated utility service, 6, 36
Relationship:
   curvilinear, 41
   straight line, 41
   cost estimating, 40–43
   requirements, 17, 20–26
Retroactive pricing, 9, 119, 172
Risk factor:
   technical, management, and cost risk of the job, 131–133
   relatively difficulty of the work, 131–133
   size of the job, 132–133
   period of performance, 132–133
   supplier's capital investment, 132–133
   assistance by the buyer, 132–133
   amount of subcontracting, 132–133
Risk percentage, 130–133
Risk spectrum, 130
Risk:
   assessment of, 10
   technical, 10, 131–133
   management, 131–133
   cost, 10, 131–133

Salvage value, 123–127
Secondary method of price analysis, 40
Secondary reliance on cost analysis, 137. *See also* Price analysis with incremental cost analysis; Primary reliance on price analysis
Semifixed, 65–67. *See also* Semivariable
Semivariable, 65–67. *See also* Semifixed
Share ratio, 14–17, 27–29. *See also* Fee adjustment ratio
Single source, 5, 172
Skill level and mix, 96
Slope of the learning curve, 98. *See also* Rate of learning
Sole source, 5
Special, export packaging, 139
Spreadsheet program, 87
Strategic Cost Analysis, 1, 3–5, 165–168, 172
Supplier cost proposal, 86
Suppliers unwilling (or unable) to provide complete cost detail, 143

Target cost, 14–17, 27–29
Target price, 14–17
Target profit, 14–17. *See also* Target Fee
Term, 17. *See also* Indefinite-delivery type; Open end
Tertiary price analysis technique, 46. *See also* Auxiliary price analysis technique
Total cost per unit, 66–72
Total cost, 66–72
Total fixed cost per unit, 66–72
Total fixed cost, 66–72
Total landed cost, 138–143
Total variable cost, 66–72
Transactional analysis, 3–8, 172
Truth-in-Negotiations Act, 86. *See also* Cost or pricing data; Current, accurate, and complete; Federal norm
Type of contract, 9–32

U.S. Army Corps of Engineers, 118
U.S. Department of Labor Bureau of Labor Statistics (BLS), 38
U.S. port of entry, 139
Unallowable cost, 9
Uniform Commercial Code, 142
Unit learning curve, 97. *See also* Cumulative average (learning) curve
Unknown-unknown, 142. *See also* Known-known and Known-unknown
Useful life, 124–127
Utility qualities, 46. *See also* Aesthetic qualities; Value Analysis

Value analysis, 46. *See also* Aesthetic qualities; Utility Qualities
Variable cost per unit, 66–72
Variable cost, 64–65, 172
Vertical integration, 167
Visual analysis, 46–47

Weighted average wage rate analysis, 112–115
Weighted profit percentage, 130–133
Wharfage, 139